INERA Workshop: Transition Metal Oxide Thin Films – Functional Layers in "Smart Windows" and Water Splitting Devices 2014

Parallel Session of the 18th International School on Condensed Matter Physics

Journal of Physics: Conference Series Volume 559

Varna, Bulgaria
4 – 6 September 2014

Editors:

T. Ivanova
K. Gesheva
H. Chamatti
G. Popkirov

ISBN: 978-1-63439-668-4
ISSN: 1742-6588

Printed from e-media with permission by:

Curran Associates, Inc.
57 Morehouse Lane
Red Hook, NY 12571

Printed by Curran Associates, Inc. (2015)

For permission requests, please contact the Institute of Physics at the address below.

Institute of Physics
Dirac House, Temple Back
Bristol BS1 6BE UK

Phone: 44 1 17 929 7481
Fax: 44 1 17 920 0979

techtracking@iop.org

Additional copies of this publication are available from:

Curran Associates, Inc.
57 Morehouse Lane
Red Hook, NY 12571 USA
Phone: 845-758-0400
Fax: 845-758-2634
Email: curran@proceedings.com
Web: www.proceedings.com

TABLE OF CONTENTS

Journal of Physics: Conference Series **559** (2014) 011001 doi:10.1088/1742-6596/559/1/011001

Preface

The Special issue presents the papers for the INERA Workshop entitled "Transition Metal Oxides as Functional Layers in Smart windows and Water Splitting Devices", which was held in Varna, St. Konstantin and Elena, Bulgaria, from the 4th-6th September 2014.

The Workshop is organized within the context of the INERA "Research and Innovation Capacity Strengthening of ISSP-BAS in Multifunctional Nanostructures", FP7 Project REGPOT 316309 program, European project of the Institute of Solid State Physics at the Bulgarian Academy of Sciences.

There were 42 participants at the workshop, 16 from Sweden, Germany, Romania and Hungary, 11 invited lecturers, and 28 young participants. There were researchers present from prestigious European laboratories which are leaders in the field of transition metal oxide thin film technologies. The event contributed to training young researchers in innovative thin film technologies, as well as thin films characterization techniques. The topics of the Workshop cover the field of technology and investigation of thin oxide films as functional layers in "Smart windows" and "Water splitting" devices. The topics are related to the application of novel technologies for the preparation of transition metal oxide films and the modification of chromogenic properties towards the improvement of electrochromic and termochromic device parameters for possible industrial deployment.

The Workshop addressed the following topics:

Metal oxide films-functional layers in energy efficient devices;
Photocatalysts and chemical sensing;
Novel thin film technologies and applications;
Methods of thin films characterizations;

From the 37 abstracts sent, 21 manuscripts were written and later refereed.

We appreciate the comments from all the referees, and we are grateful for their valuable contributions.

Guest Editors:

Assoc. Prof. Dr.Tatyana Ivanova
Prof. DSc Kostadinka Gesheva
Prof. DSc Hassan Chamatti
Assoc. Prof. Dr. Georgi Popkirov

Workshop Organizing Committee

Prof.DSc Kostadinka Gesheva, Central Laboratory of Solar Energy and New Energy Sources, Bulgarian Academy of Sciences (CL SENES-BAS) - Chairperson
Assoc. Prof. Dr Anna Szekeres – Institute of Solid State Physics- BAS
Assoc. Prof Dr. Tatyana Ivanova – CL SENES -BAS
Assist. Prof. Radostina Kamburova – ISSP-BAS

Peer review statement

All papers published in this volume of *Journal of Physics: Conference Series* have been peer reviewed through processes administered by the proceedings Editors. Reviews were conducted by expert referees to the professional and scientific standards expected of a proceedings journal published by IOP Publishing.

Journal of Physics: Conference Series **559** (2014) 012001 doi:10.1088/1742-6596/559/1/012001

Thermochromic vanadium oxide thin films: Electronic and optical properties

G A Niklasson, S-Y Li and C G Granqvist

Department of Engineering Sciences, TheÅngström Laboratory, Uppsala University, Uppsala, Sweden

E-mail: gunnar.niklasson@angstrom.uu.se

Abstract. Vanadium dioxide, VO_2, is a widely studied thermochromic material with potential applications in energy efficient window technology. It undergoes a first-order metal-to-insulator transition, accompanied by a crystal structure transformation from monoclinic to tetragonal rutile, at a critical temperature of 68 °C. Below this temperature, VO_2 is semiconducting and infrared transmitting whereas it is metallic and infrared reflecting above the transition temperature. However, in order to achieve significant thermochromic switching, the luminous transmittance of thin films will typically be less than 50%. Here we report on recent research to improve the luminous transmittance as well as the transmittance change at the transition temperature. We systematically evaluate the effect of antireflection coatings, doping with Mg and the performance of coatings comprising thermochromic nanoparticles in a transparent matrix. The last option is shown to give the best performance and holds great promise for practical applications.

1. Introduction

Modern buildings have large windows and glass facades in order to give desired indoors–outdoors contact as well as good day-lighting. These glazings are often detrimental to the energy efficiency of the building and let in or out too much energy.This needs to be compensated for by indoor space cooling or heating, depending on the season. In particular, the need for cooling has increased dramatically in recent years. Chromogenic glazings—based on thermochromism or electrochromism—can regulate the inflow of visible light and solar energy between widely separated limits and hence achieve good energy efficiency [1]. Whereas electrochromic smart windows are beginning to appear on the market, the thermochromic option has so far been little used in practical applications. However, thermochromic smart windows have potential advantages in terms of simplicity of operation and have been intensively researched in recent years.

There are two types of thermochromic materials: inorganic and organic. The inorganic ones are transition metal compounds, but of those only vanadium dioxide, VO_2, exhibits a switching temperature in the vicinity of room temperature [2].The switching takes place in the near infrared (NIR) and infrared wavelength regions from a low temperature transmitting state to a high temperature reflecting state. There also exist a large number of organic polymer-based thermochromic materials [3]. In particular, the ligand-exchange systems may be promising for niche applications. They switch mainly in the visible wavelength range from a low-to-medium absorbing state to a highly absorbing state as the temperature increases [3].For completeness we mention also thermotropic materials [3], whose working principle is a temperature-dependent change of optical scattering.

In this paper we review recent developments in the field of inorganic thermochromic coatings with emphasis on doped vanadium dioxide coatings and VO_2 nanoparticles.

2. Optical properties of vanadium dioxide films and nanoparticles

Vanadium dioxide, VO_2, is of particular interest since its optical properties change at a temperature that is not too far from room temperature [2]. Specifically, the switching in VO_2 occurs between a low-temperature phase with semiconducting properties and a high-temperature phase with metallic-like conductivity. The switching temperature is around 68 °C and exhibits significant hysteresis. Vanadium dioxide has been attracting attention with regard to fundamental research as well as window applications for many years; see [4–6] and references therein.

A thin film of VO_2 exhibits a large NIR transmittance at low temperatures and a low transmittance at high temperatures, as illustrated in figure 1.The left-hand panels show spectral transmittance $T(\lambda)$ (upper) and reflectance $R(\lambda)$ (lower) for a 50-nm-thick VO_2 film in the wavelength range corresponding to solar radiation. It is clearly seen that $T(\lambda)$ is much larger in the semiconducting state than in the metallic state—*i.e.*, below and above the phase change temperature—for wavelengths $\lambda > 1$ μm.

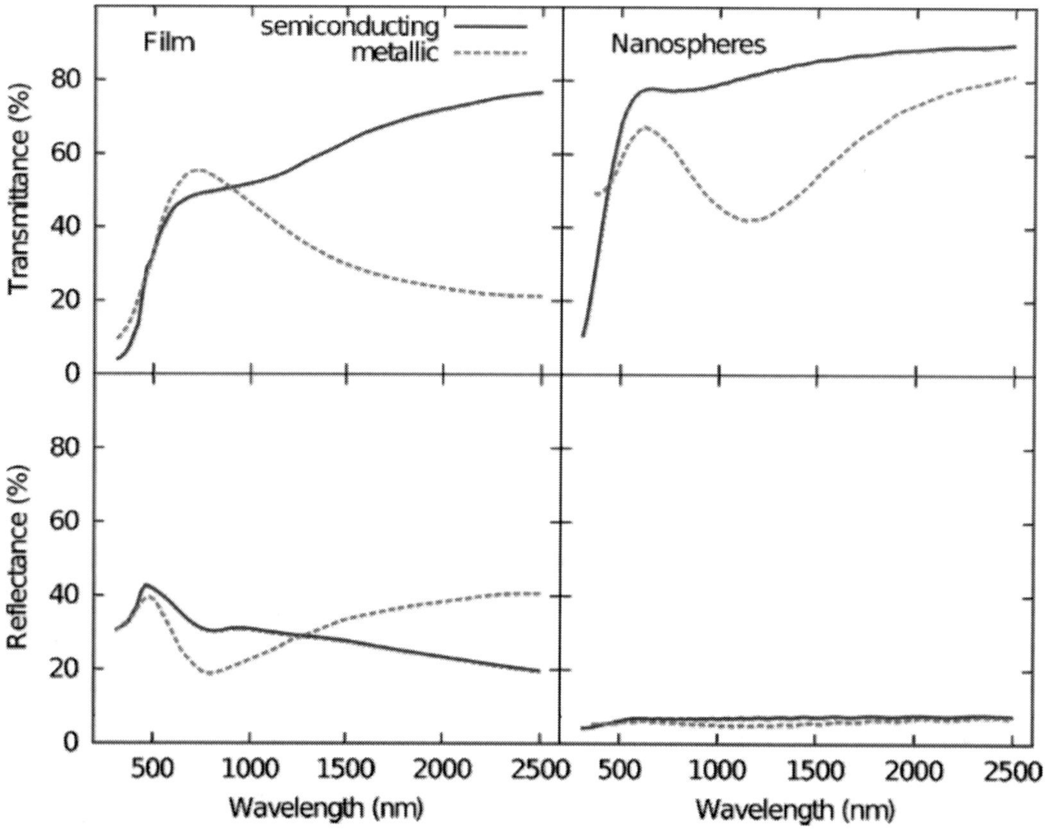

Figure 1. Spectral transmittance (upper panels) and reflectance (lower panels) for a sputter deposited 50-nm-thick thermochromic VO_2 film (left-hand panels) and as computed for a layer consisting of a dilute dispersion of VO_2nanospheres, with an equivalent VO_2 thickness of 50 nm, in a medium mimicking typical transparent polymers (right-hand panels).

The difference in $T(\lambda)$ between low and high temperature grows with increasing wavelength. It is also observed that the change in reflectance is considerably lower. However, $R(\lambda)$ increases monotonically towards long wavelengths, as expected for a metallic state. Hence a glazing with a thermochromic VO_2 film lets in significantly more energy at low temperature than at high temperature [7].

The right-hand panels in figure 1 show the computed spectral transmittance and reflectance for a layer of VO_2nanospheres in a matrix.The material was taken to be a 5-μm-thick layer of a transparent medium, with properties representative of typical polymers, containing 1 vol.% of well dispersed VO_2nanospheres with diameters much less than any relevant wavelength. This choice of parameters gives an equivalent VO_2 thickness of 50 nm, which makes the data for the nanospheres and the film straightforward to compare. It is seen that the nanoparticle composite is less reflecting and more transmitting than an equivalent thin film having the same amount of VO_2. In the metallic state, a pronounced absorption occurs around a wavelength of approximately 1200 nm. Hence the switching in the composite is from a low-temperature transmitting state to a high-temperature NIR absorbing state. The conduction band electrons in the metallic phase of VO_2 are confined within the nanoparticles and give rise to localized plasma resonance absorption. This is the physical reason for the thermochromism of VO_2 nanoparticles [8].

We now consider the implication of these properties for applications in energy-efficient fenestration. We can identify three major challenges for practical implementation of thin films of VO_2 on glazings [8]:

(*i*) The shift between the semiconducting and metallic states takes place at a switching temperature τ_c of about 68°C, which obviously is too high for buildings-related applications.

(*ii*) The visible transmittance is only about 40–50 %, which is too low for most practical window applications. Of course, a film thinner than 50 nm would be more transparent, but such films have a smaller change of the solar energy transmittance between the semiconducting and metallic states.

(*iii*) The transmittance switching is strong only for wavelengths far out in the NIR, at which the solar irradiation is quite weak. This limits the solar energy modulation to below 10%, which is too small to be of much practical interest.

Issue (*i*) is straightforward to resolve. Replacing a few per cent of the vanadium atoms with tungsten to make $W_xV_{1-x}O_2$ films can decrease the transition between the semiconducting and metallic states to room temperature as illustrated in figure 2 [9]. Doping with Mg gives a similar effect, but the drop of the transition temperature is smaller in this case.

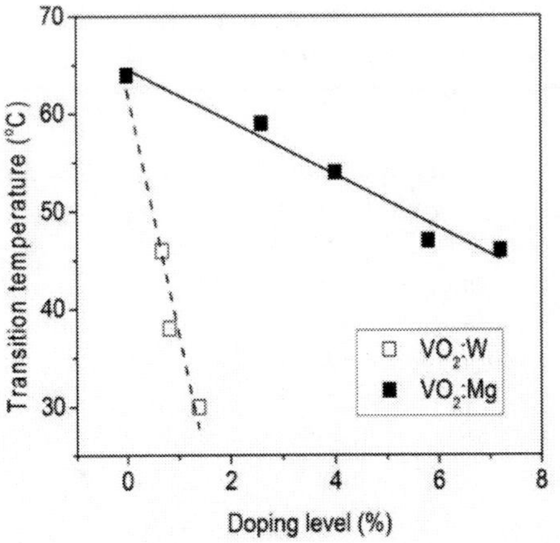

Figure 2. Temperature for the transition between semiconducting and metallic states in thermochromic thin films of $W_xV_{1-x}O_2$ and $Mg_yV_{1-y}O_2$, where x and y are referred to as "doping level" and lines were drawn for convenience. From [9].

Issue (*ii*) is associated with optical absorption at short wavelengths, which is clearly seen in figure 1. This absorption is due to the fact that VO_2 has a rather small optical band gap, of the order of 1.6 to 1.8 eV [10], between the *O2p* and π^* bands [11]. This band gap can be made larger by adding a dopant, for example magnesium. The addition of Mg displaces the transmittance edge at ~500 nm towards shorter wavelengths, but it also decreases the transmittance change between the semiconducting and metallic states at long wavelengths [9]. It was shown that the luminous transmittance goes from 39 to 51% when the Mg content is increased from zero to 7.2% [9].The short wavelength transmittance shift is indeed due to a widening of the optical band gap, as clarified recently for $Mg_yV_{1-y}O_2$[10]. However, widening of the fundamental band gap does not completely solve the problem of low visible transmittance. Vanadium dioxide also exhibits a smaller band gap, of the order of 0.5 eV, between a split-off $Vd_{||}$ band and the π^* band [11]. Electronic transitions between these two bands give rise to a low optical absorption in the wavelength range between the two band gaps, which ultimately limits the visible and NIR transmittance. However, it is seen in figure 1 that the material containing nanospheres is considerably more transparent than the corresponding film.

It can be realized from figure 1 that the limited modulation of the solar energy transmittance between the semiconducting and metallic states in VO_2 can be improved by using a layer containing VO_2 nanoparticles rather than a continuous VO_2 film. The optical switching takes place at shorter wavelengths in the nanocomposite than in the homogeneous film. The main part of the switching occurs between 700 and 1500 nm in the case of a nanocomposite, which is a region where the solar NIR radiation has an appreciable intensity. The localized plasma resonance is almost ideally located for solar control. Calculations have shown that the solar energy modulation can be increased up to 20% [12,13]. In addition, the particle size should be of the order of 20 nm or less in order not to produce too large optical scattering ("haze") [14].

3. Experiments and calculations

3.1. Sample preparation and characterization

Thin films of Mg-doped and pure VO_2 were prepared on glass and carbon substrates by reactive DC magnetron sputtering in a deposition system based on a Balzers UTT 400 unit. Co-sputtering was performed from 5-cm-diameter targets of vanadium (99.5%) and Mg (99.9%) onto substrates kept at ~450 °C. More than forty samples were grown to thicknesses in the range $40<d<$ 100nm at a rate of ~0.06 nm/s.

The value of *d* was measured with a Bruker Dektak XT profilometer and was further confirmed with optical analysis.

The elemental compositions of samples deposited on carbon substrates were evaluated using Rutherford backscattering spectrometry (RBS) with 2 MeV ^4He ions backscattered at an angle of 170°. Data were then extracted from analyses based on iterative least-square fitting to experimental spectra by use of the SIMNRA program [15]. More detailed information on the experimental procedures has been presented elsewhere [16].

Spectral and temperature-dependent normal transmittance $T(\lambda,\tau)$ and near-normal reflectance from the surface side $R_s(\lambda,\tau)$ and back side $R_b(\lambda,\tau)$ of the samples were measured using a single-beam spectrophotometer devised for absolute measurements and a Perkin–Elmer Lambda 900 double-beam spectrophotometer equipped with a $BaSO_4$-coated integrating sphere.

Data were recorded in the wavelength range $300 <\lambda< 2500$ nm; measurements were performedat room temperature and $\tau \approx 100$ °C. Optical constants were obtained by simulating the complex dielectric functions $\varepsilon = \varepsilon_1 + i\varepsilon_2$ with oscillator models and fitting the experimental data on R_s, R_b and T to calculated spectra using commercial software [17]. The optical constants for a large number of VO_2 and VO_2:Mg films were subsequently used to compute the performance limits of thermochromic thin films and nanoparticle composites, as detailed below.

3.2. Calculations of optical properties

Calculations based on Fresnel's equations [18] were used to derive $T(\lambda,\tau)$ for VO_2-based thin films with thicknesses $0 < d < 200$ nm and backed by glass plates characterized by a refractive index 1.5. The input data were the optical constants for all our samples.

In order to model the optical properties of nanocomposite layers, we must first obtain the effective dielectric function of the nanocomposite. The modelling considers VO_2 nanoparticles embedded in a polymer-like dielectric medium characterized by the dielectric constant $\varepsilon_m = 2.25$ (refractive index 1.5). The optical properties of such a composite layer can be described with effective medium theory, assuming nanoparticle sizes that are much smaller than the wavelength. In the same manner as in earlier work [12], we derive an effective dielectric function ε^{MG} from Maxwell–Garnett theory [19] according to

$$\varepsilon^{MG} = \varepsilon_m \frac{1 + \frac{2}{3} f\alpha}{1 - \frac{1}{3} f\alpha}, \tag{1}$$

where f represents the filling factor (*i.e.*, volume fraction) of VO_2 particles and is taken to be 0.01 for the dilute material of present interest. The parameter α is given for spherical particles by

$$\alpha = \frac{\varepsilon_p - \varepsilon_m}{\varepsilon_m + \frac{1}{3}(\varepsilon_p - \varepsilon_m)}, \tag{2}$$

where ε_p is the dielectric function of the VO_2 particles. Finally ε^{MG} for VO_2-polymer composites with different Mg contents was used as an input for computations of $T(\lambda,\tau)$ and $R(\lambda,\tau)$. These calculations used Fresnel's equations in the same way as for thin films. The layer thickness for the nanocomposites was varied in the 0 – 20 μm range to enable comparison with data for films containing the same amount of VO_2-based material and shown above.

3.3. Performance criteria

The performance of VO_2 thin films and nanoparticle composite layers were characterized in terms of their wavelength-integrated luminous and solar transmittance. These properties were obtained from

$$T_{\mathrm{lum,sol}}(\tau) = \int d\lambda \varphi_{\mathrm{lum,sol}}(\lambda)\; T(\lambda,\tau) \;/ \int d\lambda \varphi_{\mathrm{lum,sol}}(\lambda)\;, \tag{3}$$

Where φ_{lum} is the spectral sensitivity of the light-adapted human eye [20] and φ_{sol} is the solar irradiance spectrum for air mass 1.5 (the sun at 37° above the horizon) [21]. The thermochromic performance, and its relevance for energy efficient fenestration, was specified by T_{lum} for $\tau < \tau_c$ and $\tau > \tau_c$, T_{sol} for $\tau < \tau_c$ and $\tau > \tau_c$, and ΔT_{sol} defined by

$$\Delta T_{\mathrm{sol}} \equiv T_{\mathrm{sol}}(\tau < \tau_c) - T_{\mathrm{sol}}(\tau > \tau_c). \tag{4}$$

In the following, we will primarily use the parameters T_{lum} and ΔT_{sol} to compare the performance of thermochromic thin films and nanocomposites.

4. Results and performance limits

Figures 3 and 4 display computed data on the luminous transmittance in the low-temperature semiconducting state and on ΔT_{sol}, respectively, as a function of Mg content. Data for VO_2:Mg thin films with thickness 50 nm and for VO_2: Mg nanocomposites having an equal amount of vanadium based oxide are compared side-by-side. It is seen that the luminous transmittance is significantly higher for the nanocomposites than for the homogeneous films. An increase of T_{lum} due to the addition of Mg is observed for both films and nanocomposites. Qualitatively similar results were obtained for the metallic phase of VO_2 as well [16]. In addition, figure 4 shows that $\Delta T_{sol}$is limited to approximately 0.07 for the VO_2 films but can reach as much as 0.2 for the nanocomposites. This parameter exhibits the highest values for pure VO_2 and is decreasing with increasing Mg content. Overall, the nanocomposites display much improved luminous transmittance and solar energy modulation as compared to homogeneous VO_2 thin films.

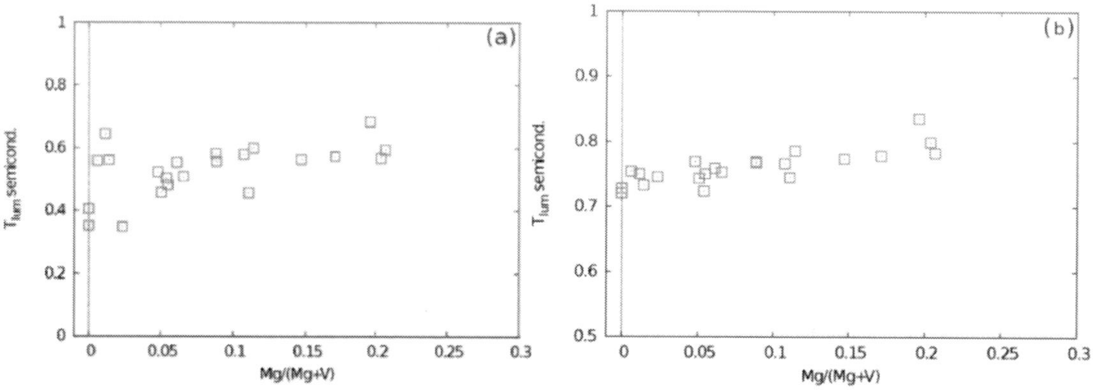

Figure 3. Luminous transmittance T_{lum} for 50-nm-thick films of undoped and Mg-doped thermochromic VO_2 (panel a) and for for 5-μm-thick nanocomposite layers containing 1 vol.% of undoped and Mg-doped thermochromic VO_2 (panel b) as a function of Mg/(Mg+V) atom ratios. From [16].

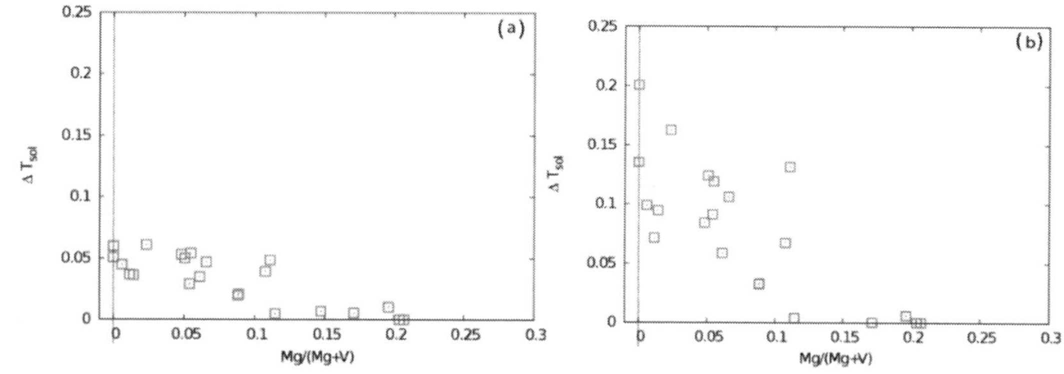

Figure 4. Solar energy transmittance modulation ΔT_{sol} for 50-nm-thick films of undoped and Mg-doped thermochromic VO_2 (panel a) and for 5-μm-thick nanocomposite layers containing 1 vol.% of undoped and Mg-doped thermochromic VO_2(panel b) as a function of Mg/(Mg+V) atom ratios. From [16].

It is evident that various combinations of luminous transmittance and solar energy modulation can be accomplished in different VO_2-based materials. Figure 5 (from ref. [16]) presents an overview over the current state-of-the-art. This figure is based on a much more detailed set of computations than those presented above, as well as on published experimental data. Pure VO_2 films give rather low values for both T_{lum} and ΔT_{sol}, but the situation can be improved by antireflection (AR) coatings. Particularly good properties have been obtained with multilayers of TiO_2 and VO_2 [22, 23]. Mg-containing VO_2 films, especially those with a low amount of Mg, have superior properties, which can be boosted by antireflection. However, the best performance is achieved with VO_2 nanoparticles which, for example, can display ~60% luminous transmittance combined with more than 20% solar energy modulation. Interestingly, Mg-containing nanoparticles do not yield better performance than nanoparticles of pure VO_2. The reason for this seems to be that the Mg partly decreases the solar modulation, as shown above, to an extent that is not compensated by the better transmittance at shorter wavelengths.

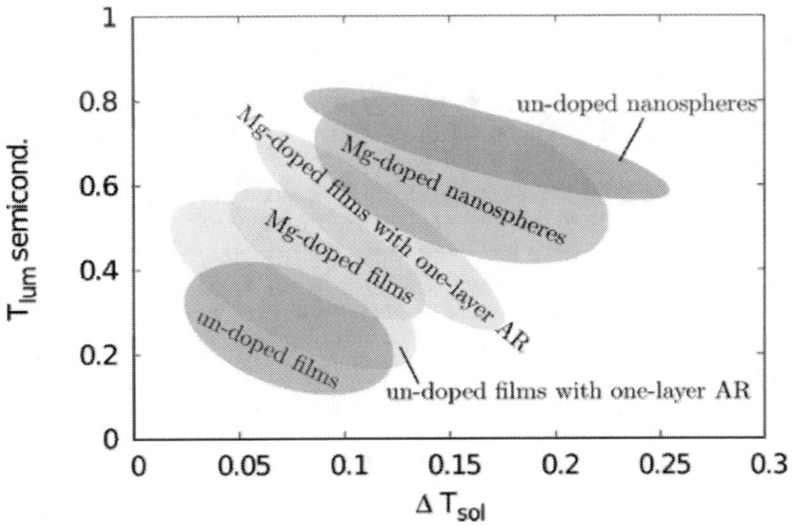

Figure 5. Schematic illustration of performance limits of the shown VO_2-based materials with regard to window applications. Specifically shown are luminous transmittance T_{lum} for semiconducting films and particles, and solar energy transmittance modulation ΔT_{sol} for thin films—including antireflection (AR) coated ones—and dilute nanoparticle composites. From [16].

5. Conclusions

We have given a comprehensive review on thermochromc VO_2-based materials with regard to applications in energy efficient windows. The optical constants of sputter deposited undoped and Mg-doped VO_2 films were used in model computations for optical properties of undoped and Mg-doped VO_2 films and dilute nanoparticle composite layers with different thicknesses. It was found from computational results that Mg-doped VO_2 films with $Mg/(Mg + V) < 0.06$ and judiciously chosen thicknesses can have much improved luminous transmittance and somewhat improved solar energy modulation compared to pure VO_2 films. Modelling of nanoparticle composites suggested that undoped VO_2 nanoparticles are more favourable than Mg-doped ones. The nanocomposites displayed superior values of both luminous transmittance and solar energy modulation and are of prime interest for applications in energy-efficient smart windows. The technological challenges related to the large-scale production of VO_2-based nanocomposites are great, though. VO_2 nanoparticles can be produced by a variety of chemical and physical methods but they must be incorporated into a passivating matrix material to prevent oxidation of the particles. The possibility to produce VO_2 particles [24, 25] and composites [26] by sputtering is particularly interesting in this respect, since sputtering is a potentially upscalable technology.

Acknowledgements
Financial support was received from the Swedish Research Council and from the European Research Council under the European Community's Seventh Framework Program (FP7/2007–2013)/ERC Grant Agreement No. 267234 (GRINDOOR).This paper was presented at INERA Workshop "Transition Metal Oxide Thin Films: Functional Layers in 'Smart Windows' and Water Splitting Devices", September 4–6th, 2014, Varna, Bulgaria. The Workshop is part of the Program of INERA REGPOT Project of Institute of Solid State Physics, Bulgarian Academy of Sciences.

References
[1] Smith G B and Granqvist C G 2010*Green Nanotechnology: Solutions for Sustainability and Energy in the Built Environment* (CRC Press, Boca Raton, FL, USA), Chapter 4
[2] Jorgenson G V and Lee J C 1986 *Sol. Energy Mater.***14** 205–214
[3] Seeboth A, Ruhmann R and Muhling O 2010 *Materials***3**5143–5168
[4] Babulanam S M, Eriksson T S, Niklasson G A and Granqvist C G 1987 *Sol. Energy Mater. Sol. Cells***16** 347–363
[5] Tomczak J M and Biermann S 2009 *Phys. Status Solidi B* **246** 1996–2005
[6] Warwick M E A and Binions R 2014 *J. Mater. Chem. A***2** 3275–3292
[7] Saeli M, Piccirillo C, Parkin I P, Binions R and Ridley I 2010 *Energy Buildings* **42** 1666–1673
[8] Li S-Y, Niklasson G A and Granqvist C G 2012 *Thin Solid Films* **520** 3823–3828
[9] Mlyuka N R, Niklasson G A and Granqvist C G 2009 *Appl. Phys. Lett.* **94** 171909
[10] Li S-Y, Mlyuka N R, Primetzhofer D, Hallén A, Possnert G, Niklasson G A and Granqvist C G 2013 *Appl. Phys. Lett.* **103** 161907
[11] Hu S, Li S-Y, Ahuja R, Granqvist C G, Hermansson K, Niklasson G A and Scheicher R 2012 *Appl. Phys. Lett.* **101** 201902
[12] Li S-Y, Niklasson G A and Granqvist C G 2010 *J. Appl. Phys.* **108** 063525
[13] Li S-Y, Niklasson G A and Granqvist C G 2011 *J. Appl. Phys.* **109** 113515
[14] Laaksonen K, Li S-Y, Puisto S R, Rostedt N K J, Ala-Nissila T, Granqvist C G, Nieminen R M and Niklasson G A 2014 *Sol. Energy Mater. Sol. Cells* **130** 132–137
[15] Mayer M 1999 *AIP Conf. Proc.* **475** 541–544
[16] Li S-Y, Niklasson G A and Granqvist C G 2014 *J. Appl. Phys.* **115** 053515
[17] Theiss M 2002 *Hard and Software for Optical Spectroscopy* (Aachen, Germany); http://www.mtheiss.com.
[18] Pfrommer P, Lomas K J, Seale C and Kupke C 1995 *Solar Energy* **54** 287-299
[19] Maxwell–Garnett J C 1904 *Philos. Trans. R. Soc. London, Ser. A* **203** 385-420
[20] Wyszecki G and Stiles W S 2000 *Color Science: Concepts and Methods, Quantitative Data and Formulae*, 2nd ed. (Wiley, New York, USA, 2000).
[21] ASTM G173-03 Standard Tables of Reference Solar Spectral Irradiances: Direct Normal and Hemispherical on a 37° Tilted Surface, *Annual Book of ASTM Standards* (American Society for Testing and Materials, Philadelphia, PA, USA, 2008), Vol. 14.04; http://rredc.nrel.gov/solar/spectra/ am1.5.
[22] Mlyuka N R, Niklasson G A and Granqvist C G 2009 *Phys. Status Solidi* A **206** 2155–2160
[23] Chen Z, Gao Y, Kang L, Du J, Zhang Z, Luo H, Miao H and Tan G 2009 *Sol. Energy Mater. Sol. Cells* **95** 2677–2684
[24] Clavero C, Slack J L and Anders A 2013 *J. Phys. D Appl. Phys.* **46** 362001
[25] Montero J (2014), private communication
[26] Romanyuk E, Steiner R, Marot L, Spassov V and Oelhafen P 2008 *Thin Solid Films* **516** 8513–8516.

IOP Publishing
doi:10.1088/1742-6596/559/1/012002

APCVD Transition Metal Oxides - Functional Layers in "Smart windows"

K A Gesheva[1], T M Ivanova and G K Bodurov
Central Laboratory of Solar Energy and New Energy Sources, Bulgarian Academy of Sciences, Blvd. Tzarigradsko chaussee 72, Sofia, Bulgaria

E-mail: kagesh@phys.bas.bg

Abstract. Transition metal oxides (TMO) exhibit electrochromic effect. Under a small voltage they change their optical transmittance from transparent to collored (absorbing) state. The individual material can manifest its electrochromic properties only when it is part of electrochromic (EC) multilayer system. Smart window is controlling the energy of solar flux entering the building or car and makes the interiors comfortable and energy utilization more effective. Recently the efforts of material researchers in this field are directed to price decreasing. APCVD technology is considered as promising as this process permits flow-through large-scale production process. The paper presents results on device optimization based on WO_3-MoO_3 working electrode. Extensive research reveals that WO_3-MoO_3 structure combines positive features of single oxides: excellent electrochromic performance of WO_3 and better kinetic properties of MoO_3 deposition. The achieved color efficiency of APCVD WO_3-MoO_3 films is 200cm^2/C and optical modulation of 65-70% are practically favorable electrochromic characteristics. To respond to low cost requirement, the expensive hexacarbonyl can be replaced with acetylacetonate. We have started with this precursor to fabricate mixed $W_xV_{1-x}O_3$ films. The films possess excellent surface coverage and high growth-rate. CVD deposition of VO_2, a promising thermochromic thin film material is also presented.

1. Introduction

Recently, the necessity for global environmental conservation and the reduced use of electrical energy and decrease of CO_2 have led to the development of energy-saving systems. Electrochromic devices (ECDs) and especially "Smart windows" as energy efficiency-related have recently attracted considerable attention because of their low switching voltage, the ability to display various colors, high reflective contrast ratio, large viewing angle, memory effect and long-term functioning. The properties of transition metal oxides, serving as functional layers in ECD are mainly determined by their specific band structure and electron distribution. They possess d-electrons, which create d-band partially overlapping s-band. The d-band provokes the interesting chemical, physical and optical properties of transition metal oxides. Electrochromism is defined as effect of color change caused by applied voltage and has been found in several different metal oxides, the metals belonging to transition series. This experimental fact suggests close relation between electrochromic properties and electronic structure of the functional material.

[1] Corresponding author

Major challenges in the development of smart windows involve the low cost fabrication of large-area devices with high optical quality. The electrochromic phenomenon was first discovered in tungsten oxide (WO_3) [1], which is still considered to be one of the best choices for the primary working electrode in EC devices [2] and remains the most promising candidate for large-scale EC devices [3]. EC effect occurs in the intercalated thin films of the smart window (WO_3) by dual injection (cathodic) or ejection (anodic) of ions and electrons. A typical reaction for a cathodic colouring material using lithium ions is [4]:

$$WO_3\ (\text{colorless}) + yLi^+ + ye^- \leftrightarrow Li_yWO_3\ (\text{blue})$$

To assure this process a multilayer system is needed, functional electrode, counter electrode, and electrolyte between assuring ions in the system. Electrochromic films as working electrodes must be conductive for ions and electrons and after ions insertion from electrolyte or from an adjacent ion conductor there will be a charge-balancing counter flow of electrons from the transparent conductors. These electrons will evoke the modulation of optical properties and will remain in the EC film as long as the ions stay there. Ion conductor (electrolyte) can be a thin film or a bulk-like polymeric material or a liquid. Its role is to let a quick transport of ions between the EC film and the counter electrode without electrons passing through. The ion conductor should have high ion conductivity and negligible electronic conductivity, and it must be transparent in the range $0.35\ \mu m \leq \lambda \leq 1.5\ \mu m$ if the device is for architecture applications. Counter electrode is an important part of the EC device. It must be conductive for ions, as well as electrons. In order to have efficient counter electrode, it must meet some requirements: necessity of reversible preservation of charge under the form of injected ions and serve as ion storage; to be transparent in the spectral range of the device application; under the influence of ion injection to be optically passive or to exhibit electrochromism in a sense opposite to that of working EC electrode and to be long-term stable when cycling.

The basic characteristics of electrochromic device "Smart windows" as practical window are contrast ratio (CR) at least 4:1; coloring and bleaching times (switching speed) of a few minutes; operating glass surface temperatures of 20°C to 80°C; switching with applied voltage of 1-5 V; open circuit memory of a few hours (maintains a fixed state of transmission without corrective voltage pulses); acceptable neutral color; large area with excellent optical clarity; sustained performance over 20-30 yr; acceptable cost ($\$100/m^2$).

Major challenges in the development of smart windows involve the low cost fabrication of large-area devices with high optical quality. One important part of the EC price is due to the expensive transparent conductive coating. Polymeric flexible substrate materials are under development to replace the conductive glass [5]. Recent results [6] are related to a new transparent conducting layer namely ZnO doped with gallium (GZO) and other doped ZnO films.

WO_3 was studied in details in our group [7, 8] and it shows excellent electrochromic characteristics. The problem with APCVD WO_3 is that the growth-rate is too small at low temperatures. We accepted 200-300°C deposition temperature, because at higher temperatures, the alkali elements in the glass substrates diffuse, and if temperature gets over 400°C the glass starts "softening". Keeping low temperatures when glass is involved is very important. We started to investigate APCVD MoO_3 expecting to grow faster at lower temperature. MoO_3 exists in many polymorph phases, one of which, the thermodynamically stable layered structure phase of MoO_3, often marked as α – phase is with an orthorhombic symmetry. The internal interaction among atoms is dominated by ionic and covalent forces, the sub layers pair interact by the weak van der Vaals forces [9]. Molybdenum oxide is known as a cathodic electrochromic material since long ago [10].

Molybdenum oxide films were obtained by atmospheric pressure chemical vapor deposition (APCVD) method, using the molybdenum carbonyl as a vapor precursor in an oxygen rich ambient. Molybdenum hexacarbonyl is a white powder which at room temperature does not react with air, with very low temperatures of sublimation and its powder density is $Mo(CO)_6 = 1.96\ g/cm^3$. V. Syijrkin [11] presents a formula for determination of vapor pressure at a given temperature:

$$\log p = A - \frac{B}{T}$$

where $\boldsymbol{p}$ – vapor pressure is in units кPa, and $\boldsymbol{A}$ and $\boldsymbol{B}$ are constants.

Most probably the chemical reaction that takes place in the CVD reactor can be schematized as:

$$Mo(CO)_6 + O_2 \rightarrow MoO_3 + CO_2\uparrow$$

We have experimented on different types of substrates and varying technological conditions [12-13]. The structure and film properties are strongly influenced by the technological process parameters, a large number of experiments were done for obtaining the films with optimal optical quality. The experimental approach is fixing one parameter and varying the others. The altered technological parameters of APCVD process are the substrate temperature ($T_{deposition}$), the sublimator temperature ($T_{sublimator}$) and the gas ratio ($Mo(CO)_6/Ar$). It has been established that MoO_3 films are successfully obtained in the temperature range of 150-200°C with MoO_3 film thicknesses ranging 320-400 nm for the 150°C set of samples and 120-240 nm for the 200°C-set of samples [14]. The two sets were grown for one and the same deposition time. The lower-temperature set process was proceeding with higher deposition rate of 10 nm/min compared to 6 nm/min for the higher temperature samples, resulting in thicker films. This behavior is associated with thermal decomposition in the vapor phase.

Experiments with increasing the deposition temperatures above 200°C (300 and 400°C) showed that there was no film deposits nevertheless there were enough precursors vapors in the reactor, because powder deposits on the reactor walls were registered. This was a sign that this particular CVD process and the chosen construction of CVD equipment for MoO_3 film deposition requires lower substrate temperatures in comparison with the analogical process for WO_3 films, which were successfully deposited in the temperature range from 200 to 400°C [11, 15, 16] Previous CVD research has focused primarily on producing molybdenum oxide films from chlorides, fluorides and carbonyls as vapor precursors. Lander and Germer [17] first deposited MoO_3 by low pressure CVD (LPCVD) by using $Mo(CO)_6$. Thin films of polycrystalline orthorhombic MoO_3 films were prepared and studied by double-step CVD at atmospheric pressure [18]. J. S. Cross et al [19] proceeded a thermodynamical study of low pressure chemical vapor deposition (LPCVD) of MoO_3, where the precursor used is Mo $(CO)_6$. The substrate temperature is varied in the range of 300-450°C. Solid orthorhombic α-MoO_3 is the most stable phase over these ranges of the temperatures. T. Maryama et al. [20] proposed APCVD method using $Mo(CO)_6$ with substrate temperature varied 150 to 400°C and they obtained polycrystalline MoO_3 at 350°C. The authors [20] found that the deposition rate increased with temperature to 250°C with exhibited peak at 300°C, then a decrease for substrate temperatures above 300°C. They determined color efficiency 25.8 cm^2C^{-1} at λ=550 nm.

2. Experimental

Using our experiences in WO_3 and MoO_3 film depositions, mixed films on their basis were deposited. The precursors are the corresponding hexacarbonyls. The hexacarbonyl precursors ($W(CO)_6$, $Mo(CO)_6$ were mixed in a ratio $Mo(CO)_6$:$W(CO)_6$=1:4. The deposition proceeded in argon-oxygen rich ambient at atmospheric pressure (APCVD process) in a horizontal cold walls CVD reactor. The precursor powder placed in a sublimator immersed in silicon oil bath has been heated at temperature of 90°C and has been controlled with an accuracy of ±1°C. This sublimator temperature provides a sufficient vapor pressure of the hexacarbonyls. The gas lines ("Galtek" type Teflon) were heated up to the sublimator temperature in order to assure successful transport of the vapors to CVD reactor. Argon (99.995%) flow carries the precursor vapors to the reactor. The selected flow rate of Ar through the sublimator assures a constant amount of the precursors vapor. Through a separate line, oxygen (99.95%) enters the reactor. In the present study the ratio of flow rates of Ar/O_2 is 1:32 [21-24].

For comparative study the deposition temperature was kept at 200°C because only at that temperature it was possible for the three kinds of the metal oxide films to be obtained. WO_3 can be prepared at higher temperatures (up to 400°C) and the growth rate increases with increasing temperature. The growth rate of MoO_3 decreases with increasing temperature due to the gas phase reaction [25, 12]. From this point of view, the temperature of 200°C is a cross point for the deposition of the mixed oxide films. The deposition time duration was kept constant (40 min).

The characterization methods applied are: UV - VIS Spectrophotometry characterizing transmittance and reflectance of CVD films, estimation of optical band gap energy (Eg). Auger Electron Spectroscopy (AES) for determination of chemical composition. FTIR Spectroscopy was employed for the investigation of vibrational properties and the chemical bonds; Raman Spectroscopy for determination of crystallization degree, phases and chemical bonds. Scanning Electron Microscopy (SEM), Transmission Electron Microscopy (TEM) and Atomic Force Microscopy (AFM), field emission scanning electron microscopy (FESEM) – all techniques for study of film morphology; Spectral Ellipsometry (SE) - study of the spectral dependences of optical constants (refractive index and extinction coefficient), analysis of dispersion curves of optical absorption. Cyclic Voltammetry - for investigation of electrochemical behavior of thin films under applied voltage and reversibility of electrochromic effect.

3. Results and Discussions

Optical transmittance was measured for the single oxides and the mixed oxide films. The results are shown on figure 1. It is seen that after annealing the transmittance of single oxides increases, while for the mixed WO_3-MO_3 it decreases. A possible explanation is mass transfer starting during annealing at elevated temperatures, but the process is not accomplished (annealing time was accepted to be one hour), the structure gets disordered, amorphous-like and the transmittance drops down.

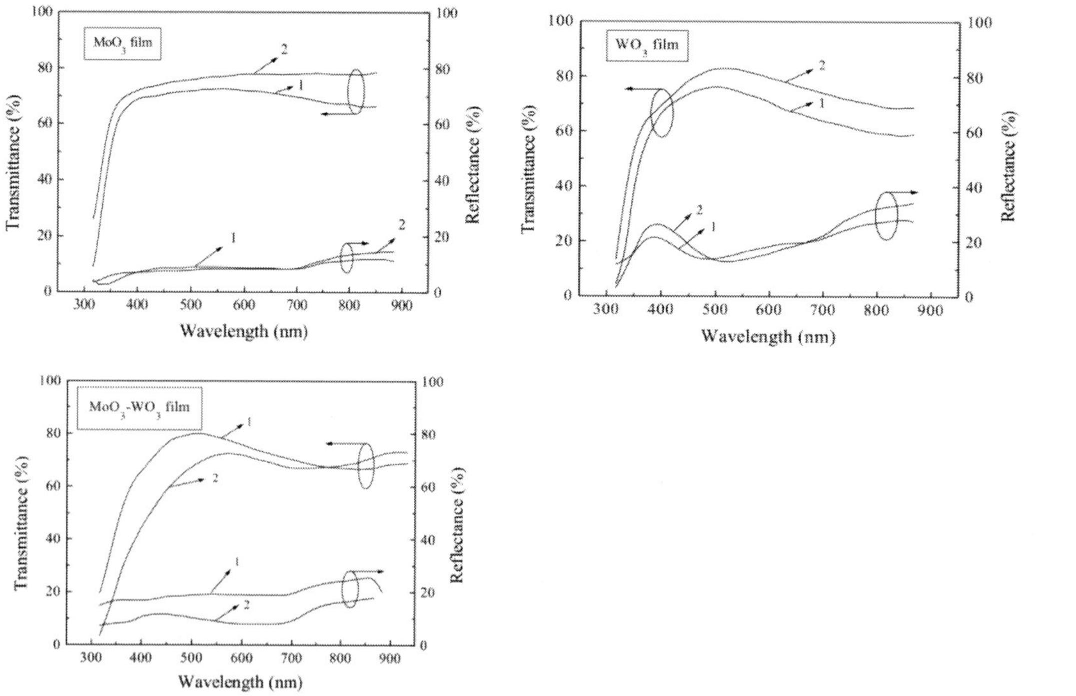

Figure 1. UV-VIS spectra of single oxides and mixed in as-deposited (1) and annealed (2) at 400°C. See the lower transmittance of mixed films after annealing

The IR spectra recorded for the three types of films after 500°C annealing, have presented the different behavior once again of MoO_3 and the main band located at 834 cm^{-1} (figure 2). This absorption band is related to Mo - O - Mo vibrations. The IR bands of $W_xMo_{1-x}O_3$ and WO_3 are broad with greater overlapping of the two spectra with three maxima observed at 810, 717 and 640 cm^{-1}. The bands at 810 and 640 cm^{-1} can be assigned to stretching W-O vibrations of crystalline monoclinic WO_3 and the absorption line at 717 cm^{-1} is due to bending W=O vibrations. The IR spectra reveal that in the structure of $Mo_xW_{1-x}O_3$ films prevails the tungsten oxide fraction. The shifting of the main absorption band for the mixed film is another evidence for the presence of a new phase.

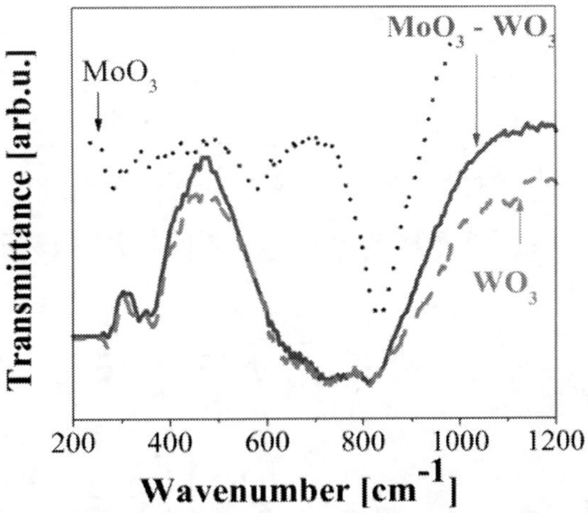

Figure 2. IR spectra of the 500°C annealed films of MoO_3, WO_3 and MoO_3-WO_3.

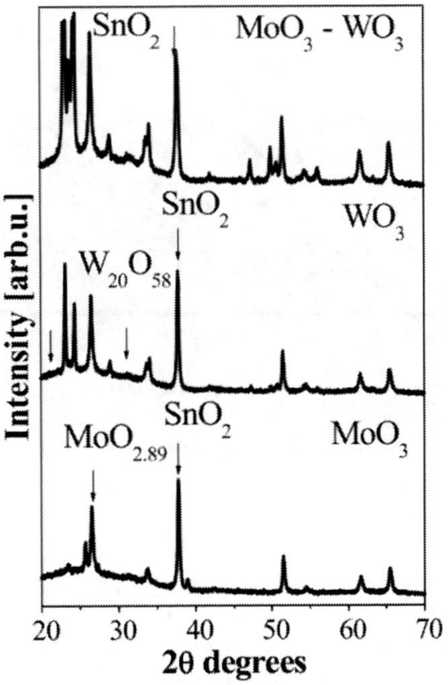

Figure 3. XRD spectra of MoO_3, WO_3 and MoO_3 -WO_3 films, deposited on conductive glass substrates at the deposition temperature of 200°C and annealed at 400°C.

XRD patterns of MoO_3, WO_3 and MoO_3 -WO_3 films, deposited on conductive glass substrates are presented on figure 3. Most of the XRD lines found can be assigned to the corresponding oxides. MoO_3 film consists of orthorhombic Mo trioxide and a small fraction of sub-oxide. The triclinic WO_3 phase is prevailed in the tungsten oxide film with $W_{20}O_{58}$ sub-oxide. The XRD spectrum of the mixed oxide film is very similar in shape to that of WO_3. Similar effect is observed for IR spectra. The XRD peak positions are the same with enhanced intensity, which leads to suggestion that W atoms are partly substituted by Mo atoms.

Figure 4 presents high resolution TEM micrograph of CVD as-deposited MoO_3-WO_3 film, grown at gas flow ratio 1/32. The columnar structure of the film is clearly observed, it is typical for the CVD grown films at temperatures below the melting point of the film material. These columns form grain boundaries which facilitate the ions intercalation.

AFM study reveals grained structure of CVD grown films. As seen from figure 5, there is a contrast change over a grain, which is evidence for clusterization in the film structure. These grains also facilitates the ions interaction, the coefficient of diffusion is bigger. The average surface roughness for MoO_3 is 31,4 nm, for WO_3 is 24,8 nm and for mixed oxide film is 45,7 nm, respectively. It must be noted that the film thickness is different, although the deposition parameters are the same.

Figure 4. High resolution TEM micrograph of CVD as-deposited MoO_3-WO_3 film, grown at gas flow ratio 1/32.

Figure 5. AFM images of annealed at 300°C CVD MoO_3 and WO_3 and mixed Mo - W oxide films (the precursor is mixture of Mo and W carbonyl powders).

Cyclic voltammetry is used as method to study insertion reactions and to receive information about the electrochemical effects. Cyclic voltammetric experiments were performed in a standard three-electrode arrangement. The electrodes were immersed in electrolytes of 1 mol/l $LiClO_4$ in propylene carbonate (PC). Current density vs. voltage voltammograms were registered between -1 V and +1.5 V. The cyclic voltammetric (CV) curves manifest that the as-deposited and annealed MoO_3-WO_3 films showed considerably a larger current density compared to the those of the single metal oxide (MoO_3, WO_3) films (see figure 6).

There is an improvement in the electrochromic properties with annealing. However, the most interesting feature is the rapid increase in the electrical charge (see table 1) for the mixed oxide films, proving its suitability and quality for applying in EC devices. The inserted and extracted charge density had been estimated from CV curves and they showed that the exchanged charges for WO_3-MoO_3 are much higher than those for single oxides.

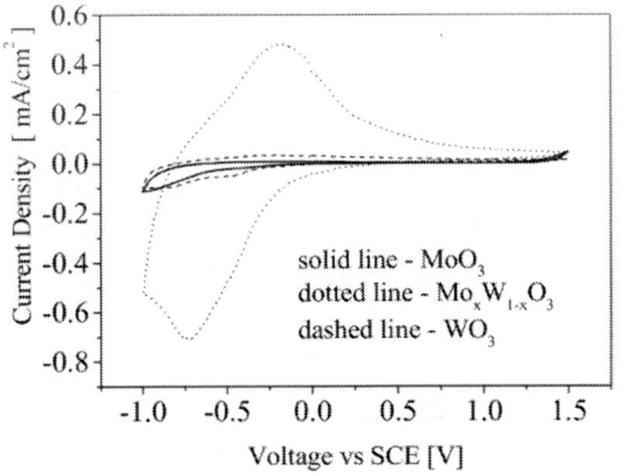

Figure 6. Cyclovoltametric curves of CVD MoO_3, WO_3 and MoO_3-WO_3 films, annealed at 400°C for 1 hour.

Table 1. Inserted and extracted charge density determined from CV data.

CVD Films		$Q_{inserted}$ [mC/cm²]	$Q_{extracted}$ [mC/cm²]
MoO_3	As deposited	31.57	30.36
	Ann. 400°C	16.62	41.18
WO_3	As deposited	23.56	20.53
	Ann. 400°C	30.20	31.40
MoO_3-WO_3	As deposited	116.3	170.9
	Ann. 400°C	195.59	204.5

Electrochromic characteristics were determined from the voltamograms. Color efficiency can be determined from:

$$CE = \frac{\Delta(OD)}{\Delta Q_{ins}},$$

where the optical density (OD) is related to extinction and film thickness:

$$OD = (\frac{4\pi}{\lambda})kd,$$

Experimentally, the change of optical density Δ(OD) can be determined from the relation:

$$\Delta(OD) = -\log 10(\frac{T_{colored}}{T_{bleached}}),$$

Journal of Physics: Conference Series **559** (2014) 012002 doi:10.1088/1742-6596/559/1/012002

where $T_{colored}$ is the transmittance measured after ion injection and $T_{bleached}$ – transmittance of bleached film (extraction of alkali ions).

Color efficiency values for the three kinds of films, annealed at 500°C have been calculated and the results for different wavelengths are shown on figure 7. The color efficiency (CE) strongly depends on the wavelength and the changes of the optical density ΔOD. The CE values in the visible range for MoO_3-WO_3 films are better than those for WO_3 films.

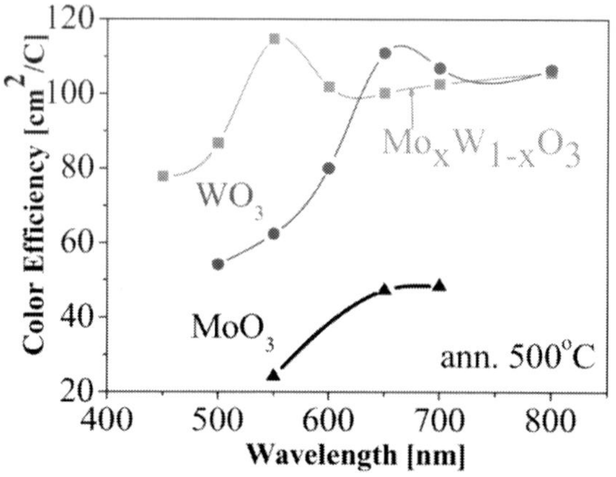

Figure 7. Color efficiency values of CVD MoO_3, WO_3 and MoO_3-WO_3 films after annealing at 500°C for 1 hour in air

Table 2 presents the estimated values of color efficiency and optical modulation of Mo-W oxides, depending on the scan-rate of the potential applied. Optimization is made related to the scan-rate of the applied potential by changing it as follows: 2, 5, 10, 20 and 50 mV/s.

Table 2. Electrocromic characteristics of MoO_3-WO_3 films, obtained from CV curves, measured at different scan rates.

Scan rate [mV/s]	**I_{anodic} [mA]**	**$I_{cathodic}$ [mA]**	**Δ Q [mC]**	**ΔT (550 nm)**	**CE [cm²/C] (550 nm)**
50	1.22	-1.68	19.18	21.10	43,49
20	0.64	-0.76	20.91	29.96	53.97
10	0.36	-1.08	30. 64	45,21	196,46
5	0.09	-0.63	33.54	49.59	88.69
2	0.08	-0.39	65.90	21.20	43.49

Optimization is made related to the scan-rate of the applied potential by changing it as follows: 2, 5, 10, 20 and 50 mV/s. Table 2 presents the estimated values of color efficiency and optical modulation of mixed Mo-W oxide films depending on the scan-rate of the potential applied.

The high values of the color efficiency are obtained for mixed WO_3-MoO_3 films and after optimization it had been found that the optimal potential scan-rate is 10mV/s. The optical modulation, namely the difference between the transmittance in colored and bleach states was also estimated, and the results are seen in table 2. The best values (estimated for 550 nm, maximum of the solar spectrum) at the optimal potential scan-rate is 45%.

Two kinds of EC devices were prepared, integrated mixed WO_3-MoO_3 films in as-deposited state and after 500°C annealing. The solid electrolyte was prepared with 1 mol/l $LiClO_4$ in PC and adding polyvinyl alcohol: WO_3-MoO_3 /$LiClO_4$+PC+PVA/SnO_2: Sb. Table 3 shows that EC device involving as deposited mixed oxide films showed lower transmittance after intercalation of Li ions, better electrochromic effect.

Table 3. Transmittance values before and after ion intercalation.

	Initial Transmittance	Transmittance after Li injection
WO_3-MoO_3, As deposited	48.5 %	17.4%
WO_3-MoO_3, 500°C annealed	46.0%	29.0%

Other interesting electrochromic materials are vanadium oxide and mixed W-V-O films. VO_2 undergoes a phase change across a critical temperature of 68°C. It is a promising thermochromic material for energy efficient fenestration [25]. Vanadium dioxide is largely studied and we try to add to the common knowledge results on APCVD carbonyl process deposited VO_2 films. Recently, we deposited thin films employing a mixed precursor of $W(CO)_6$ and $V(CO)_6$ [26]. However utilizing $V(CO)_6$ is not reasonable as it is too expensive. Thus, we adopted the cheaper vanadyl acetylacetonate as possible vanadium precursor. The tungsten hexacarbonyl has lower sublimation temperature, it would supply more vapors at higher temperature compare to acetylacetonate. To avoid two separate vapor source installations, we succeeded to obtain mixed oxide films employing one vapor source approach. We started with CVD deposition of vanadium oxide films first to establish the lowest possible deposition temperature, and vapor source temperature [27, 28].

The thickness of pure VO_2 films deposited at 400°C and vapor source temperature of 170°C (first set of samples) is 349 nm; refractive index 1.53. After annealing at 500°C for 1 hour, the thickness remained the same, 350 nm closed to as deposited VO_2 film. VO_2 films deposited at 400°C substrate temperature, and 140°C vapor source temperature (second set of samples) has thickness of 316 nm.

FTIR spectra, shown on figure 7 of the two sets of samples revealed the following: from 400 to 750 cm^{-1} the bands are mainly attributed to the stretching modes of doubly coordinated oxygen (V_2–O) and triply coordinated oxygen (V_3–O) bonds. The peak at 414 cm^{-1} is attributed to stretching V-O vibrations [29]. The peak at 538 cm^{-1} is attributed to the bridged V–O–V stretching vibrations (longer V–O bonds). The peak at 554 cm^{-1} corresponds to the stretching vibrations of the (V-O-V) bridging bonds. The peaks at 697, 809 and 812 cm^{-1} are due to the lattice vibrations in vanadium oxide network [30]. The peaks at 865 and 907 cm^{-1} are attributed to VO_2 (B) IR bands [30].

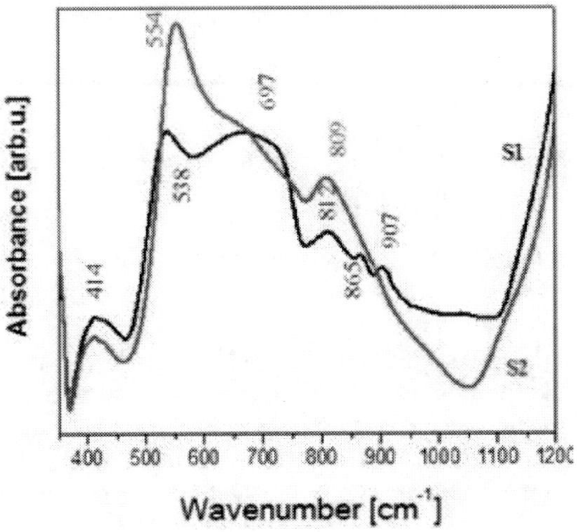

Figure 7. FTIR spectra of the two VO_2 samples (as-deposited and annealed at 500°C).

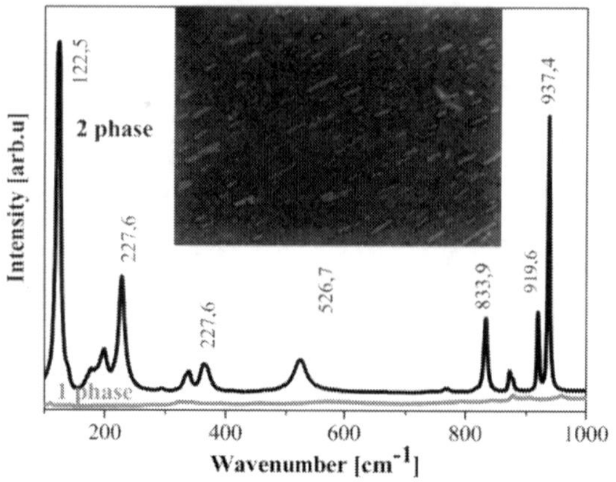

Figure 8. Raman spectra of CVD VO_x film after 500°C annealing. The film is consisted of two phases: amorphous (phase 1) and crystalline (phase 2), the embedded image reveals the two phases as the monocrystals give the crystalline phase.

Raman spectra shown in figure 8 reveal the existence of two phases – amorphous-like and polycrystalline. Very strong and clear peak is observed at 122.5 cm^{-1}, which is assigned to the stretching mode of $(V_2O_2)_n$, which in turn corresponds to the chain translation [30]. The weak low frequency Raman peak at 198.5 cm^{-1} is due to VO_2, corresponding to lattice vibrations. This peak is strongly associated with the layered structure [31]. The peaks centered at 293,9 , 411,9, 526,7 cm^{-1} are related to stretching and bending modes of V_2O_5. The weak peak at 816,2 cm^{-1} is assigned to the B_g modes of VO_2 [30].

VO_2 films deposited at substrate temperature of 400°C are thick, well adhered and with two separate phases observed on film surface: well oriented in one direction crystals and amorphous phase. It can be due to the chemical reaction nature. The high mobility necessary for stronger orientation at lower deposition temperatures most probably is ensured by the enthalpy of the ongoing chemical processes on the films surface.

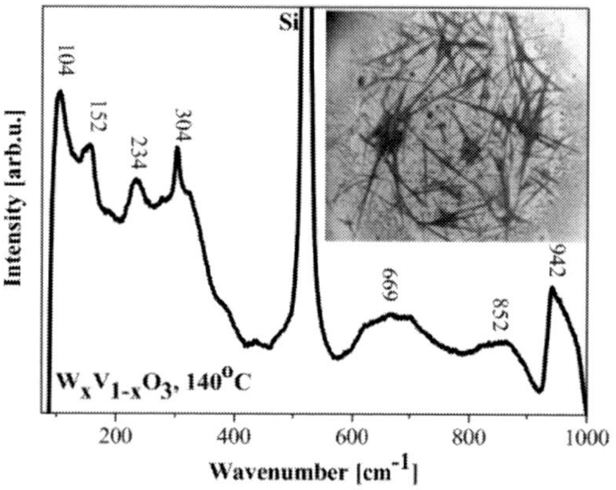

Figure 9. Raman spectrum of $W_xV_{1-x}O_3$ film on Si, the intensity axis is extended in order for the Raman Lines to be seen.

The structure and optical properties of CVD-mixed oxides, deposited from precursors mixture $W(CO)_6 : V(CO)_6 = 10:1$ are also studied. The thickness of W-V-O film deposited at 550°C and vapor source temperature of 140°C is 198 nm and refractive index is 1.96. Thickness decreases at higher deposition temperature (550°C), which is a sign that mixed vapors are seeking lower deposition temperature. The mixed W-V- O films looked like porous, visually separate islands-like-shapes were seen as randomly distributed through the film surface (see figure 9). There is a predominant fraction of amorphous phase. The strongest peak at 520 cm^{-1} is related to Si substrate and it is sharp and high. The peak at 104 cm^{-1} is due to lattice vibration. Raman peaks at 145 and 197 cm^{-1} are associated with layered structure of vanadium oxide [33]. In our case this peak is shifted towards 152 cm^{-1}. The weak peak at 234 cm^{-1} is related to δ(O-W-O) bonds. The band at 304 cm^{-1} is attributed to bending vibrations of crystalline WO_3. The strong peak around 525 cm^{-1} is related to the Si substrate. The peak at 669 cm^{-1} is related to double coordinated oxygen V^{2+}-O stretching mode or monoclinic WO_3. The broad band at 852 cm^{-1} and the peak at 942 cm^{-1} are related to V^{4+}-O bonds and W=O. The results are cross-proved by FTIR measurements, which are not included here. On the films surface from the same set of mixed W-V-O oxide films deposited on glass substrate well seen are spherical "droplets-like" particles evenly distributed over the surface area of the glass.

On figure 10 the voltamograms for as-deposited and annealed W-V-O films are presented. No significant change is observed in the behaviour of the switching functional film after annealing at 500°C. Table 4 contains data for the colour efficiency (CE) and the optical modulation. The transmittance after annealing at 500°C W-V-O films is higher than for the as-deposited films (data not shown here). Nevertheless, the calculated value for the colour efficiency is lower, which means that the intercalated electrical charge is changed in a way that the resulting value for the CE is decreased.

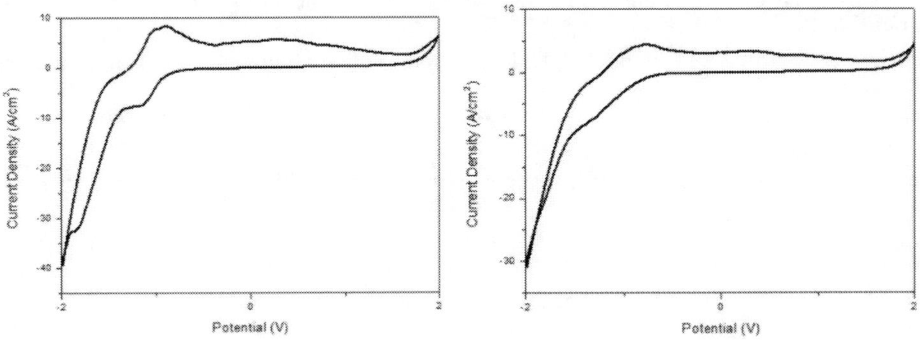

Figure 10. CV curves for as-deposited and annealed at 500°C W-V-O films.

Table 4. Electrochromic characteristics (colour efficiency and optical modulation) of mixed W-V-O thin films as determined from voltammetric measurements at λ=550 nm.

WO_3-V_2O_5	CE [cm^2/C]		Optical modulation [%]	
Precursor 10:1; Ar/O_2=1/32	As dep.	Ann. 500°C	As dep.	Ann. 500°C
	20.11	11.04	65.2	73.5

At the same time, the value of the optical modulation, namely the difference between the transmittance in colored and bleached state is increased from 65% up to 73-74%. Depending on the application envisaged, where either the CE or optical modulation is more important, the stage of film annealing can either be kept or removed from the overall technology of W-V-O electrochromic films.

4. Conclusions

The APCVD technology is a suitable deposition method for preparation of transition metal oxide thin films to serve as successful functional layers in electrochromic devices. The mixed WO_3/MO_3 films are found to combine the desirable good properties of the molybdenum and tungsten oxides. Mo/W oxide films express high color efficiencies and optical modulation approaching the electrochromic properties of the most studied EC material – WO_3. Mixed oxide films produced by APCVD method grow faster comparing to single APCVD WO_3 films.

The results from the tested electrochromic cells of transition metal oxides based on tungsten, molybdenum and vanadium are a proof for applicability of the APCVD films as functional electrodes in electrochromic devices for smart windows and displays.

Acknowledgements
This paper was presented at INERA Workshop "Transition Metal Oxide Thin Films: Functional Layers in "Smart windows" and "Water Splitting devices", September 4-6th, 2014, Varna, Bulgaria. The Workshop is part of the Program of INERA REGPOT Project of Institute of Solid State Physics, Bulgarian Academy of Sciences

References
[1] Deb S K 1969 *App. Opt.* Suppl. **3** 192
[2] Cui H N, Costa M F, Teixeira V, Porqueras I and Bertran E 2003 *Surf. Sci.* **532** 112
[3] Granqvist C G1999 *Electroch. Acta* **44** 3005
[4] Lampert C M 1988 *Solar Energy Mater Solar Cells* **52** 207
[5] Aleksandrova M, Dobrikov G, Gesheva K, Bodurov G, Cholakova I and Kolev G 2013 *Inter. J. Reasoning-Based Intelligent Systems* **5** 246
[6] Chan Ch, Tang Ch, Lee K H, Jaing C C, Kuo C, Chen H and Lee C 2010 *SPIE* **7786** 77860S-6
[7] Gogova D, Gesheva K and Stoyanov G 1994 *SPIE* **2255** 332
[8] Gogova D, Iossifova A., Ivanova T, Dimitrova Z and Gesheva K 1999 *J. Crystal Growth* **198-199** 1230
[9] Fukushima C, Nagano M, Sumita T, Kubota H, Nagata M, Honda Y, Oku T and Imahori J 1997 *Physica B* **239** 56
[10] Tubbs M R 1974 *Phys. Stat. Solidi (a)* **21** 213
[11] Syijrkin V G 1984 *Metal Carbonyls*, Chimii, Moscow, (in Russian)
[12] Gesheva K, Ivanova T, Iosifova T, Gogova D and Porat R 1999 *J. de Physique IV France* **9** Pr 8-453
[13] Gesheva K, Ivanova T, Szekeres A, Maksimov A and Zaitzev S 2001 *J. Phys. IV France* **11** Pr3 - 1023
[14] Ivanova T, Surtchev M and Gesheva K 2002 *Materials Letters* **53** 250
[15] Gogova D, Gesheva K, Kakanakova-Georgieva A and Surtchev M 2000 *EPJ Applied Physics* **11** 167
[16] Gogova D, Stoyanov G and Gesheva K 1996 *Renewable Energy* **8** 546
[17] Lander J J and Germer H 1948 *Trans. AIME* **175** 648
[18] Abdellaoui A, Leveque G, Donnadieu A, Bath A and Bouchikhi B 1997 *Thin Solid Films* **304** 39
[19] Cross J S and Schrader G L 1997 *Thin Solid Films* **259** 5
[20] Maryama T and Kanagawa T 1995 *J. Electroch. Soc.* **142** 1644
[21] Ivanova T, Gesheva K, Szekeres A, Maksimov A and Zaitzev S 2001 *J. Phys. IV France* **11** Pr3 - 385
[22] Szekeres A, Ivanova T and Gesheva K 2002 *J. Solid State Electroch.* **7** 17
[23] Gesheva K, Szekeres A and Ivanova T 2003 *Solar Energy Mater. Solar Cells* **76** 563
[24] Gesheva K A, Ivanova T, Kovalchuk A, Gurtowoi V and Trofimov O 2003 *Electroch. Soc. Proc.* **2003-08 (1)**, Eds. M.D. Allendorf, F. Maury, F. Teyssandier, 394
[25] Bodurov G, Ivanova T, Abrashev M and Gesheva K 2012 *J. Physics: Conference Series* **398** 012016
[26] Bodurov G, Ivanova T, Abrashev M, Nenova Z and Gesheva K 2013 *Physics Procedia* **46** 127
[27] Bodurov G, Ivanova T and Gesheva K 2013 *Phys. Procedia* **46** 149
[28] Luo Zh, Wu Zh, Xu X, Wang T and Jiang Y 2011 *Thin Solid Films* **519** 6203
[29] Nam S-P, Noh H-J, Lee S-J and Lee Y-H 2010 *Mater. Res. Bull.* **45** 291
[30] Rama N and Ramachandra Rao M S 2010 *Solid State Commun.* **150** 1041

Oxide coatings on flexible substrates for electrochromic applications

M Aleksandrova[1]
[1] Technical university of Sofia, Department of Microelectronics, "Kliment Ohridski" blvd, 8, bl. 1, zip 1000, Sofia, Bulgaria

E-mail: m_aleksandrova@tu-sofia.bg

Abstract. One of the most studied classes of materials in the modern microelectronic devices are the metal oxides. There are different metal oxide films, such as electrodes, charge injecting and electrochromic coatings for displays or "smart" windows applications. This paper aims to describe the recent achievements for oxide coating deposition for flexible electrochromic displays. Although many deposition methods for production of such films have been developed, some of the achievements in the field of RF sputtering of transparent electrodes from indium-tin oxide on low-cost polyethyleneterephthalate substrate are presented. Attention is paid on some critical issues, such as films electro-optical parameters (sheet resistance, transparency in the visible range), adhesion, degradation due to stress and patterning ability.

1. Introduction

Interest in the electrochromic devices and particularly in the electrochromic displays has increased recently, because these types of displays have the ability to modulate the light in such way that effect of "ink on paper" appears, which is favourable for the eye perception. This effect is suitable for electronic readers, watches and electronic labels – applications, requiring high readability, high contrast, and wide viewing angle. Although the electrochromic display is non-emissive, it offers all these advantages in contrast to the liquid crystal displays (LCD) [1]. Because the e-readers or labels are portable, a long battery life is required and therefore low power consumption is needed for such applications. This is one of the reasons for the efforts to be directed to the electrochromic displays due to their valuable property - memory after the power is turned off, which additionally decreases the power consumption. As it is known the electrochromic layer changes its optical density or colour as a result of injection or extraction of ions and electrons, respectively from electrolyte material and from transparent conductors, causing redox reaction. The optical state can return back to initial condition if the voltage is reversed. The threshold electrical potential necessary for activation of the process is within the range of 1.5-4 V. Some of the most studied electrochromic materials are WO_3 [2], MoO_3 [3], V_2O_5 [4], Nb_2O_5 [5], NiO [6] and their blends [7,8]. To observe the image generated in the display structure, the electrode film should be highly conductive and transparent for the visible light. Typical materials, fulfilling these requirements are tin oxide (SnO), indium tin oxide (In_2O_3:Sn) and zinc oxide (ZnO – pure and doped by boron) [9,10]. As can be seen, the oxide coatings are one of the main components in the electrochromic devices. Flexible substrates are the basic for the next generation display panels, making them more compact, lighter, thinner, and giving the opportunity to be rolled

[1] Corresponding author Mariya Aleksandrova, e-mail: m_aleksandrova@tu-sofia.bg

and brought in the pocket like a conventional sheets of paper [11]. However, there are still some challenges related to the technological development and some obstacles related to the mechanical durability during folding, that should be overcome. The main technological restriction is that the deposition process should be low temperature due to the lower melting point of the flexible substrates in comparison to the glass. From other side the process temperature is related to the film's microstructure and therefore to its electrophysical properties. Second aspect is the mechanical stress, induced in the films at bending, leading to their degradation and decreasing the long-term reliability of the device [12].

This paper aims to describe some of the latest materials and technologies for oxide film deposition on flexible substrates with its application in electrochromic devices. Results from vacuum sputtering of indium tin oxide as a transparent electrode on polyethylene terephthalate (PET) foil are presented more detailed. Most of the existing papers in the field present results for excellent devices performance and only mention that the substrate is flexible, but rarely can be found extended research about the impact of the internal stress induced in the films at flexing and how it affects to the performance. Here, some results concerning this problem are presented.

2. Oxide films for the electrochromic cells: materials and technologies for deposition

2.1. Transparent electrodes

These are the films, connecting the electrocromic display with the power supply. They should have a sheet resistance lower than 30 Ω/sq and transparency for the visible light higher than 88 % [13]. The most common methods for deposition of the thin electrode films are vacuum methods (electron beam evaporation and sputtering), chemical vapour deposition, spray pyrolysis and electrochemical deposition [14]. When the substrate is flexible special attention should be paid to the deposition temperature. One approach to improve the manufacturing efficiency and not involving high temperatures could be printing of conductive and polymeric inks on polyester substrate (figure 1). The origin of the deposited material is conductive metal oxide powder, dispersed in a polymer binder (for ITO it is styrene butadiene copolymer in toluene), ensuring durability at bending [15]. Electrochromic display with ITO replaced by antimony-doped tin oxide deposited onto TiO_2 pigments was successfully produced for contrast white background electrode. Although the conductivity decreases in comparison with non doped ITO, the stability of the new coating towards aqueous electrolytes is better.

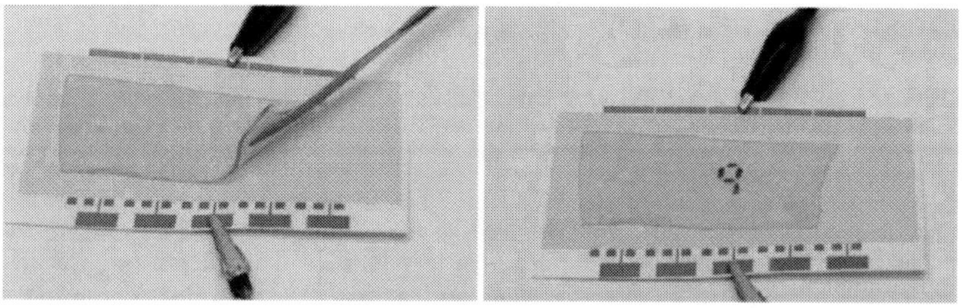

Figure 1. Seven segment printed electrochromic display [15].

For flexible display roll-to-roll sputtering a system for room temperature deposition is specially designed to produce multilayer ITO/Ag/ITO electrodes on polyethersulfone substrates [16]. The silver film thickness is negligible (less than 15 nm) in comparison to the ITO for achieving the desired front panel electrode's parameters - sheet resistance 4.28 Ω/square and optical transmittance 89.28 %. In another research, RF and DC magnetron sputtering are applied respectively for ITO (50 nm) and copper (5 nm) deposition in ITO/Cu/ITO stack on polycarbonate substrate [17]. In this way, although the transparency is decreased (~70%) the resistance is extremely small ($1.5.10^{-4}$ Ω.cm), which is a precondition for high charge injection efficiency. Multilayer structure indium zinc oxide/Ag(10nm)/indium zinc oxide(40nm)/Ag(10nm) is prepared on polyethylene terephthalate (PET) substrate by specially designed roll-to-roll DC sputtering system at room temperature and is analyzed by a bending system. This system provides bending with radius and frequency, respectively 10 mm and 60 Hz for 2 000 bending cycles [18]. Due to the silver nanoparticles existence the sheet resistance is ~5 ohm/square independently of the top indium zinc oxide film. Ag layer between both indium zinc oxide layers provides stable electrical conductivity even after the stress is applied due to the higher failure strain of the silver film. The deviation of the in-situ measured resistance, caused by the bending is less than 1 ohm/sq (20%). Aluminum-doped zinc oxide (AZO) thin films were studied as a potential candidate for flexible transparent conductive oxide polyethersulphone (PES) substrates by using an RF magnetron sputtering method at temperature below 200°C. The higher substrate temperature enhances the electric and the optical properties - the minimum electrical resistance for 250 nm thin film was 2.10^{-4} Ω.cm, and the optical transmittance was 90 % without any thermal and plasma damage to the PES substrate, which has higher degradation temperature than PET. [19]. However, the PES bending ability is weaker, due to its lower Young's modulus in comparison to the more popular PET and PEN.

Leterrier et al. observed the mechanism of revealing of cracks in pure (undopped) ITO thin films sputtered onto aromatic polyester. They found that the internal stresses were compressive, and that cracks in the ITO occur first around the pin-hole defects [20]. If the crack length is less than hundred times the thickness of the coating, it leads to increase in the electrical resistance with 10%. Otherwise, if the propagation of the crack is longer it leads to full failure of the flexible display. For evaluation of the fractional change in the resistance of ITO-coated by DC magnetron sputtering PET substrate as a function of the strain mechanical tests of the samples were made using tensile testing machine at constant crosshead speed [21]. Initial cracking occurs perpendicular to the loading direction in the ITO film at strain 2.75%. The number of cracks increases rapidly between 2.75% and 3.25% strain. At strain 6%, transverse cracking occurs due to lateral compression and at strain 10% cracks appear parallel to the primary cracks. The resistance increases sharply at threshold strain, depending on the ITO film thickness. An approach to improve the flexibility of conductive oxide electrodes on polymer substrates is introducing of surface roughness [22]. For this purpose regularly channels separated in equal distance were imprinted at elevated temperature onto Teflon® polymer surfaces. Amorphous and crystalline ITOs were magnetron sputtered. In this way ITO surfaces became unidirectionally wavy, which increased the surface area. As a result their bending stability was greatly improved - after 10,000 cycles at bend radius 10 mm, the resistance change was 50% of the conventional ITO.

2.2. Optically active materials

The most common methods for deposition of the thin electrochromic films are the same as for the transparent electrodes - vacuum methods (electron beam evaporation and sputtering), as well as electrochemical deposition, chemical vapor deposition (CVD), electrochemical oxidation of tungsten metal [11]. Again the deposition temperature is crucial not only for preventing substrate from heat damages, but for the crystallinity of the film. For example amorphous and porous tungsten oxide can be produced at substrate temperatures lower than 100°C during e-beam evaporation. During sputtering, if the temperature is revealed to 200-300°C the WO_3 films are dense and highly crystalline. Microstructure is tightly related to the electrochromic behaviour. Amorphous films have faster speed of switching between reduction and oxidation states, but for the crystalline ones the reaction is more stable and repeatable [23]. The same is valid for molybdenum oxide too, because its properties are very similar to WO_3 [24]. Recently in the literature flexible designs of multicolour electrochromic display has been reported [25,26]. Nanostructured TiO_2 layer is formed on PET by spin coating on the top of ITO electrode, using TiO_2 water dispersions mixed with polymeric dies. New developed devices work stable even at bending radius less than 1 cm. The structure has several films collored in a different way, separated by transparent electrodes (figure 2a). So here the porosity, due to low deposition temperature causes fluently ion migration, response time and voltage decrease (figure 2b). Photo of the fabricated samples can be seen in figure 3. Electrophoretic deposition process of WO_3 nanoparticles has been reported in [28]. SEM image illustrates the porosity of the film, having high surface area (figure 4). Smaller than 100-150 nm nanoparticles facilitate faster ion-insertion ability, reduce the diffusion limited effects and improve transmittance modulation from 25 to 95 %.

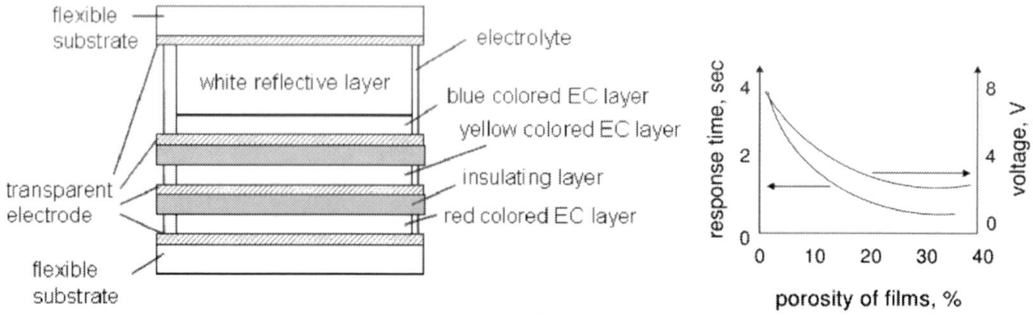

Figure 2. a) Schematic of multilayer electrochromic display; b) Response speeds and activation voltage of multilayer electrochromic display with porous films [26].

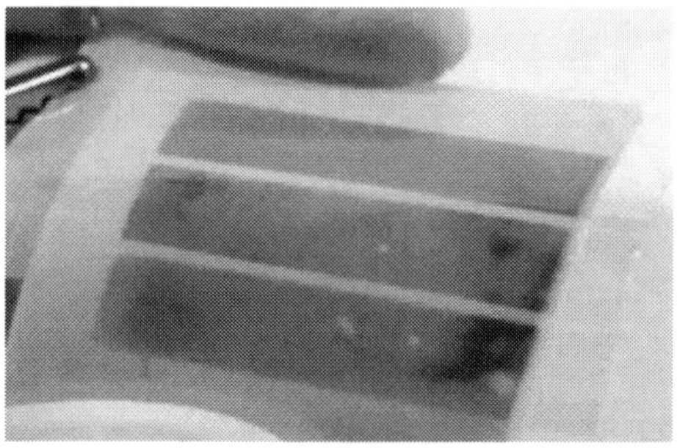

Figure 3. Three colored electrochromic display on flexible substrate [27].

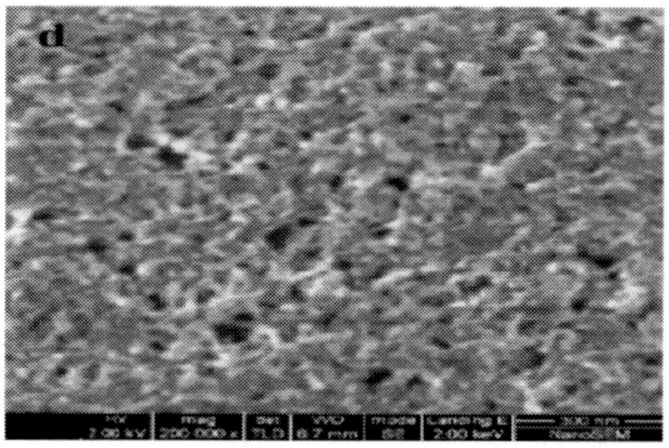

Figure 4. SEM image of a WO_3 film deposited by electrophoresis on ITO/PET [28].

Report exists about growing of plasma assisted reactively evaporated WO_3 on flexible Kapton substrate, withstanding at temperature during deposition of approximately 200°C [29]. At this temperature nanocrystalline WO_3 films are produced. Samples show 73% optical transmittance in the visible range and coloration efficiency at a wavelength 550 nm of about 45 C/cm^2. Another example presents RF magnetron reactive sputtered WO_3 on commercially available ITO coated polyethylenenaphthalate (PEN) substrate [30]. With decreasing plasma power and increasing total gas pressure, the crystalline phase disappear and the films is tend to be amorphous, which is the purpose, because such film exhibits better electrochromic properties.

3. Indium tin oxide films on PET deposited by RF sputtering –electro-optical properties and patterning ability

Although a variety of modifications has been developed in the materials and methods for adapting the transparent conductive coatings to the flexible substrates, indium tin oxide still takes one of the first places for display applications. As it can be seen from the previous section the electrode structure is rather complex in many cases (multilayers and doping) and sometimes flexibility is decreased due to using of less bendable, but thermally stable substrates. In the following text results are presented from simple, single layer deposition and tuning of the electro-optical properties of sputtered ITO films on PET by modification of the deposition regimes. PET pieces were exposed to UV light (λ=365 nm and P=250W) for 5 minutes, leading to oxidation of the polymer backbone, adsorption of oxygen species and decreased roughness of the polymer surfaces [31, 32]. In this way surface energy of the already oxidized PET was increased, indium and tin particles can be easily bonded to the surface without need of high plasma power for insuring reaction kinetic for the initial ITO monolayers. This allows deposition at temperatures lower than 100-120°C, at which PET is less thermally sensitive. Sputtering pressure in the vacuum chamber should be also precisely tuned experimentally for ensuring suitable long free paths of the metal particles, according to the target-substrate distance and sticking them to the oxygen from the PET surface. There is a minimum in the sheet resistance for an argon concentration of 25 mTorr and then it gradually increases (figure 5). The reason is decreased sputtering throughput, due to the shorter mean free path of the particles and often collisions, leading to particles' energy losses. For the optimum sputtering pressure (related to the total pressure) the minimum achieved resistivity was ~57 Ω/sq and the transparency - 87.8% (figure 6). As it can be seen from the XRD diffractogram of ITO (figure 7) its structure is mainly amorphous.

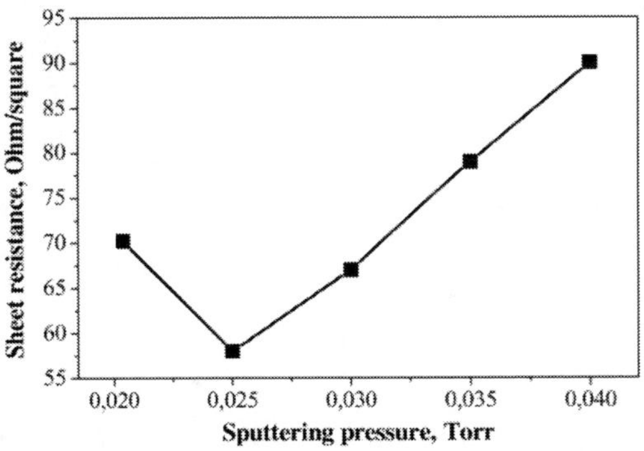

Figure 5. Sheet resistance versus sputtering pressure for ITO sputtered on PET substrate.

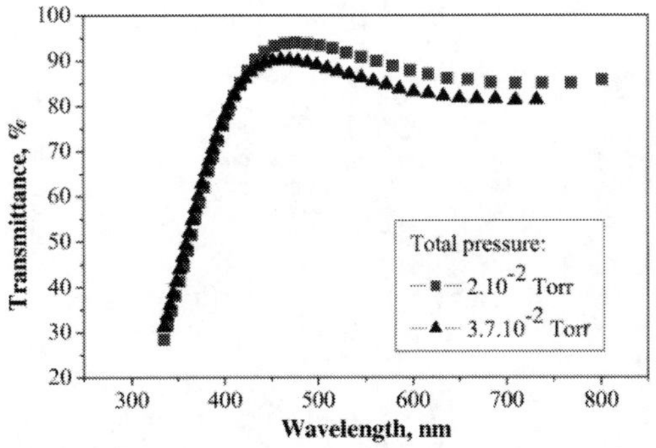

Figure 6. UV-VIS spectra of ITO coated PET at different sputtering pressures.

The strongest peak around 2θ = 30° can be ascribed to orientation of the formed crystals preferentially in (222) direction and the peaks around 45° and 55° are due to PET substrate. ITO deposition by two stage change of the plasma power on PET foil is a new contribution in the described work. The influence of the plasma intensity on the ITO film's properties was established [33,34]. This modification of the deposition mode leads to changes in the way of arrangement of the microparticles. At the first deposition stage the sputtering voltage was set to 0.5 kV (or plasma power of 75 W), although the typical values for this parameter are in the range of 1.1-1.4 kV [35]. At the lower sputtering power, the temperature in the chamber was 62°C (more than 20°C lower than the substrate's melting point). 8 minutes after the lower temperature sputtering, plasma power was increased to 105 W, corresponding to deposition temperature of 75°C. It was established by SEM (figure 8) that the power increasing leads to reorganization of the ITO particles on the substrate surface and transition from amorphous to microcrystalline structure consisting of regularly distributed grains.

Journal of Physics: Conference Series **559** (2014) 012003 doi:10.1088/1742-6596/559/1/012003

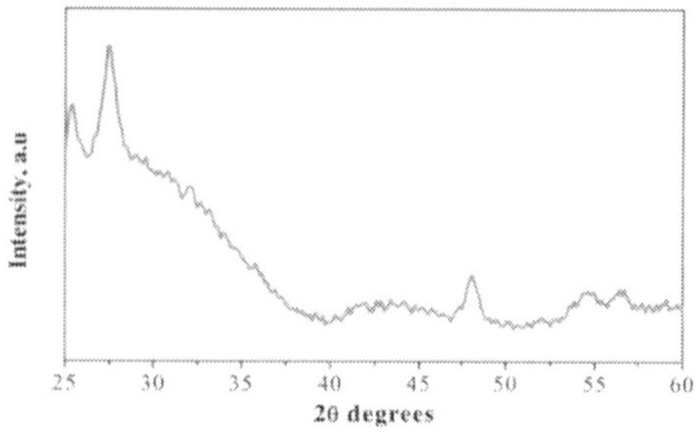

Figure 7. XRD spectrum of ITO on PET at the optimum sputtering pressure.

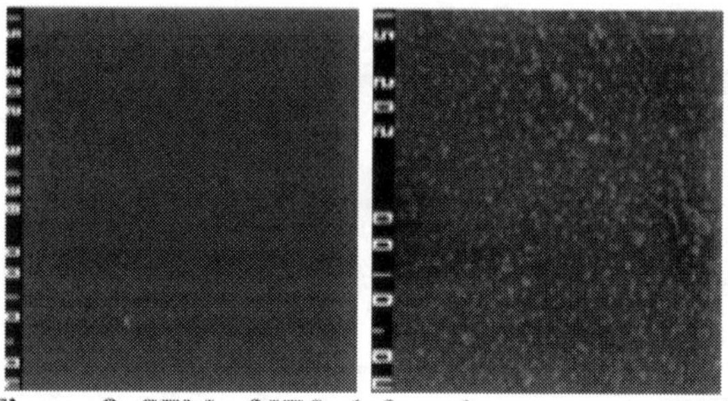

Figure 8. SEM of ITO: left – plasma power 75W; right – plasma power switched to 105W [34].

As can be seen from the left SEM image, the surface of the ITO film deposited at lower deposition temperature of 62°C is amorphous, but after plasma power increase and temperature rose to 75°C and crystallization mixed with amorphous underlayer phase presents. This structural change of the ITO film results in decreasing of the sheet resistance with 10 Ω/sq.

Two different rates of sputtering power change were set: 0.5 W/s and 3 W/s, but the power change rate of 0.5 W/s was selected as optimum, because ITO film adhesion to the PET substrate was found the strongest – at 70 N/cm^2 tensile force and peel angle of 90° the coating remained on the PET surface. At rates higher than 0.5 W/s the layer was separated from the substrate at 14 N/cm^2, because the particles suddenly receive high kinetic energy and can not arrange by taking energetically favourable states on the surface. It results in irregularities, causing strain forces and weak adhesion. Optical microscopy (figure 9) shows ITO coated PETs after cyclic mechanical loading of 500 bends with curvature ~ 135° at different sputtering power change rates. The results show few micro cracks per area of 9 cm^2 for the highly adherent ITO film, produced at sputtering power change 0.5 W/s.

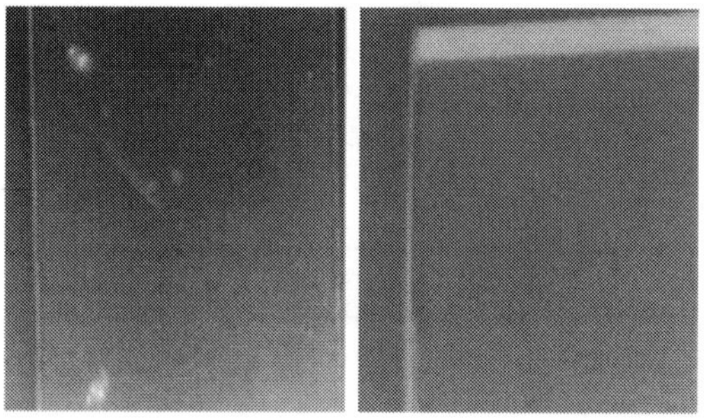

Figure 9. Optical microscopy images: left- rate of sputtering power change 3 W/s; right- rate of sputtering power change 0.5 W/s [34].

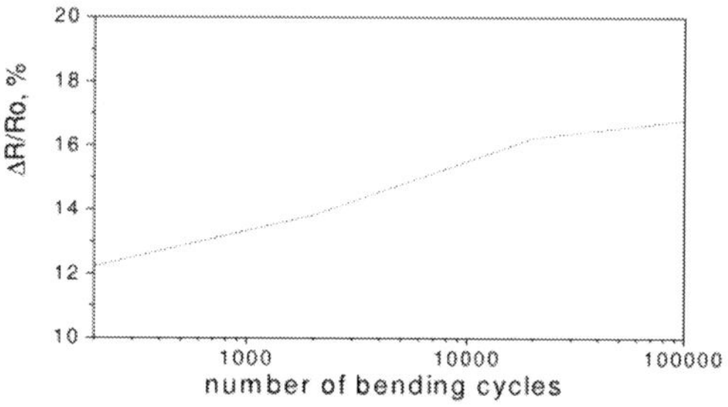

Figure 10. Relative change of the sheet resistance of sputtered ITO on PET after applying of dynamic mechanical load.

For estimation of the long term stability ITO coated PET was subjected to 100 000 numbers of mechanical bendings and the relative change of the ITO film resistance was measured. The result is illustrated in figure 10 and shows that the maximum deviation of the resistance ΔR/Ro (Ro is initial value before bending) is 17 % at 100 000 repeated bending cycles and for 2 000 cycles it is 14%. For comparison in [18], the authors report about ΔR/Ro =20% at 2 000 cycles and additionally the electrode structure is multilayer.

Post deposition UV exposure with wavelength of 365 nm produced by mercury lamp and ozone formatting was conducted for ITO work function and surface energy increase [36], which results in sheet resistance decrease (figure 11a). After 10 minutes of illumination the sheet resistance decreased to < 20 Ω/square at the same oxygen pressure. These results demonstrate that after UV treatment it is possible to increase oxygen concentration in the chamber, making the metal oxide more transparent and still conductive. Typical FTIR spectra in reflectance mode, acquired in the range 300-4000 cm^{-1} for films deposited on PET at different post-deposition conditions are shown in figure 11b. The bonds strength was not affected by the UV exposure and no breakage or degradation was indicated. The peaks are situated at the same position, which is evidence for preserving the typical features of ITO film. Exposure causes crystallites formation, indicated by X-ray diffraction [37].

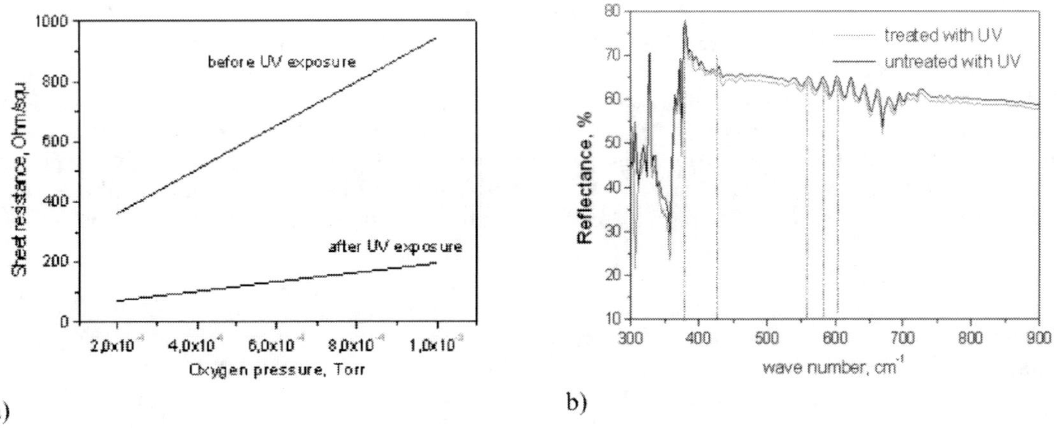

Figure 11. a) Sheet resistance of the thin ITO films at different oxygen pressures before and after 8 minutes UV exposure; b) FTIR spectra of the sputtered ITO layers on flexible PET substrate before and after UV treatment [34].

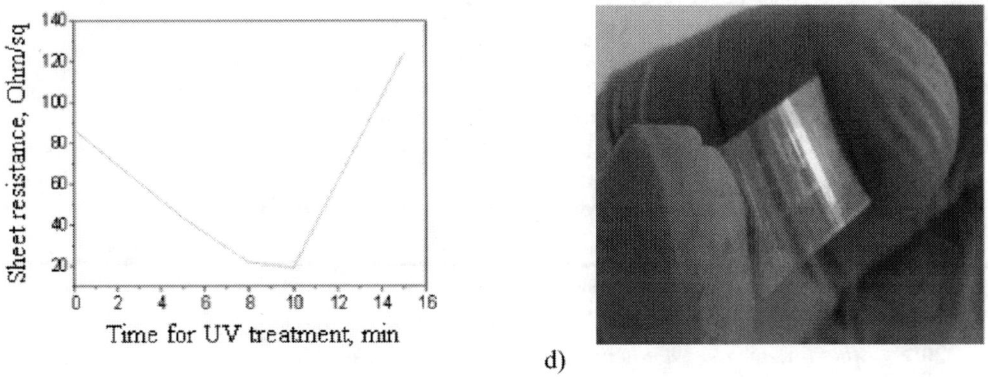

Figure 11. c) Sheet resistance of ITO sputtered on PET as a function of the UV illumination time [37]; d) ITO segments on PET, patterned by lift off process.

However, if certain critical UV dose is exceeded, degradation in the polymeric chemical bonds occurs and its resistance abruptly increases due to diffusion of released carbonyl groups into ITO film. At lower doses (shorter time than 10 minutes, 250W exposure intensity), the electrical behavior of the treated ITO layers can be ascribed to the activation of defects presenting in the disorder matter that captures charge carriers. The trapped carriers are escaped after vibrational interaction with the electromagnetic wave and the sheet and specific resistance are decreasing (figure 11c). UV illumination should be taken into account during photolithographic patterning of ITO films. For PET/ITO patterning oxalic acid solution was used. It is isotropic etcher, so undercut effect is expected. For ITO thickness of 150 nm the bias was 6 μm under both edges of the 2 mm wide strip shaped pattern – part of digital segment (figure 11 d) [38]. As alternative way for shaping of smaller patterns (such as pixels for example) lift off process is recommended, where there is no etching, so there is no bias effect.

4. Conclusion

ITO films were deposited on flexible polymer substrate by developing a modified method of RF reactive sputtering at different plasma powers. It was found that plasma power affects the ITO films crystallization and adhesion strength. The substrates were treated by UV exposure before and after ITO deposition. It was found that it affects the degree of oxidation and activation of vacation states, which results in sheet resistance improvement. The minimum value achieved was 19.2 Ω/sq., and the maximum transmittance of the visible light was above 87 %. In this way, a simple control of the electro-optical properties of sputtered ITO on PET without using of complex multilayered electrodes or doped materials is possible. The results are useful for further fabrication of flexible display.

Acknowledgements

The materials for the experiments are provided by grants DMU 03/5 and 141ПР0005-03. The used equipment is situated in department of Microelectronics, TU-Sofia. The author is thankful to colleagues helping for the measurements and to colleagues from BAS for the XRD, SEM and FTIR. The paper is presented at INERA WORKSHOP "Transition Metal Oxide Thin Films-functional Layers in "Smart Windows" and Water Splitting Devices, September 4-6 2014, Varna Bulgaria. The Workshop is part of the Program of INERA REGPOT Project of the ISSP, BAS.

References

[1] Årman S 2001 *J. New Mater. Electrochem. Syst.* **4** 173
[2] Chananonnawathorn C, Pudwat S, Horprathum M, Eiamchai P, Limnontakul P, Salawan C and Aiempanakit K 2012 *Procedia Engin.* **32** 752
[3] Dhanasankar M, Purushothaman K K and Muralidharan G 2010 *Solid State Sci.* **12** 246
[4] Yang Y, Kim D and Schmuki P 2011 *Electrochem. Commun.* **13** 1198
[5] Romero R, Dalchiele E A, Martín F, Leinen D and Ramos-Barrado J R 2009 *Sol. Energy Mater. Sol. Cells* **93** 222
[6] Bodurov G, Stefchev P, Ivanova T and Gesheva K 2014 *Mater. Lett.* **117** 270
[7] Patil CE 2014 *Curr. Appl. Phys.* **14** 389
[8] Yang Y, Kim D and Schmuki P 2011 *Electrochem. Commun.* **13** 1021
[9] Minami T 2013 *Semicond. Semimetals* **88** 159
[10] Lükermannl F, Mönkemöller V, Kurz H, Sacher M, Hamelmann F, Stiebig H and Heinzmann U 2009 *24th European Photovoltaic Solar Energy Conference, 21-25 September 2009, Hamburg, Germany*
[11] Chen J 2013 *IEEE Access* **1** 150
[12] Leterrier Y 2003 *Soc. Vac. Coat.* **505** 856
[13] Hainich R R and Bimber O 2011 *Displays: Fundamentals and Applications*, CRC Press
[14] Seshan K 2012 *Handbook of Thin Film Deposition*, Third Edition, Elsevier
[15] Coleman J P, Madhukar P and Wagenknecht J H 1999 *Sol. Energy Mater. Sol. Cells* **56** 395
[16] Park Y S, Choi K H and Kim H K 2009 *J. Phys. D: Appl. Phys* **42** 235109
[17] Kim D 2009 *Trans. Electr. Electron Mater.* **10** 165
[18] Kim H 2012 *Nanoscale Res Lett.* **7** 67

[19] Park J, Cho Y, Shin J, Cha S, Cho Ch, Kim H, Yoon S, Jeong S, Park S and Lim A 2007 *J. Korean Phys. Soc.* **51** 1968
[20] Leterrier Y, Médico L, Demarco F , Månson JAE, Betz U, Escolà MF, Kharrazi M and Atamny F 2004 *Thin Solid Films* **460** 156
[21] Cairns D R and Crawford G P 2005 *Proceedings IEEE* **93** 1451
[22] Kim J and Park J 2013 *Organic Electr.* **14** 3444
[23] Gesheva KA (Ed.) 2007 *Nova Publishers* New York
[24] Svensson J S, Benson EM and Tracy CE 1988 *SPIE proc.* **19** 1016
[25] Yashiro T 2011 *SID 11 Digest* **1** 42
[26] Yashiro T 2013 *ITE and SID* 1300
[27] Naijoh Y 2011 *ITE and SID* 375
[28] White C M, Gillaspie D T, Whitney E, Lee S and Dillon A C 2009 *Thin Solid Films* **517** 3596
[29] Krishna K H, Hussain O M and Guillen C 2008 *Optoelectron. Adv. Mat. Rapid Commun.* **2** 242
[30] Kawasaki H, Matsunaga T, Guan W, Ohshima T, Yagyu Y and Suda Y J 2009 *Plasma Fusion Res. Series* **8** 1431
[31] Helander M, Wang Z, Greiner M, Liu Z, Lian K and Lu Z 2009 *Appl. Phys. Lett.* **95** 173302
[32] Benor A, Takizawa S, Bolívar C P and Anzenbacher P 2010 *Organic Electr.* **11** 938
[33] Aleksandrova M, Dobrikov G, Cholakova I, Kolev G and Bodurov G 2012 *XI National Scientific Conference with International Participation – Electronics Proceeding* 17
[34] Aleksandrova M, Dobrikov G, Gesheva K, Bodurov G, Cholakova I and Kolev G 2013 *Inter. J.Reasoning-based Intelligent Systems* **5** 246
[35] Boiadjiev S, Dobrikov G and Rassovska M 2007 *Thin Solid Films* **515** 8465
[36] Kim B, Kim D, Jang Y, Lee N, Kwon O and Kwon Y 2007 *J.Korean Phys. Soc.* **50** 1858
[37] Aleksandrova M, Cholakova I, Bodurov G, Kolev G and Dobrikov G 2012 *J.WASET* **71** 1308
[38] Aleksandrova M, Kolev G, Cholakova I, Dobrikov G and Bodurov G 2013 *Int. J. Thin Films Sci. Technol.* **2** 67

Electrochromic Properties of Li^+-Intercalated Amorphous Tungsten ($a WO_{3-x}$) and Titanium ($a TiO_{2-x}$) Oxide Thin Films

C A Triana, C G Granqvist, and G A Niklasson

Department of Engineering Sciences, Division of Solid State Physics, The Ångström Laboratory, Uppsala University, P.O. Box 534, SE-751 21 Uppsala, Sweden

E-mail: Carlos.Triana@angstrom.uu.se

Abstract. We report on electrochromic properties of stoichiometric and oxygen-deficient amorphous films, denoted $a WO_{3-x}$ and $a TiO_{2-x}$, under Li^+-ion-electron inter/deintercalation. Optical characterization of the films in their as-deposited, fully intercalated (dark), and bleached states were performed by in-situ optical transmittance measurements. We explore electrochromism and optical absorption phenomena in the context of oxygen deficiency and nanostructure. Studies by cyclic voltammetry suggest good optical modulation and charge capacity upon Li^+-ion-electron inter/deintercalation for almost stoichiometric films.

1. Introduction

Switchable optical properties exhibited by electrochromic (EC) materials are of large importance for the development of intelligent multifunctional devices. Because of their applications in energy efficient "smart windows", photocatalysis, displays, panels for emissivity modulation and gas sensors [1,2], this type of materials has been extensively studied in the last decades [3]. Reversible optical modulation (OM) from bleached to blue-colored states in EC thin films is obtained by applying a small electric field to produce intercalation (cathodic EC) or deintercalation (anodic EC) of charge, typically electrons e^- together with small alkaline metallic ions (H^+, Li^+, Na^+, K^+) where the ion-e^- inter/deintercalation processes can occur along with oxidation phenomena [4]. Cathodic EC properties in TiO_2 films produced by several deposition techniques have been studied previously; such analyses have shown that the EC behavior in TiO_2 differs considerably among different deposition methods [5-10]. Previous results of DC sputtered crystalline TiO_2 and amorphous $a TiO_2$ films have indicated that EC behavior is strongly dependent on the deposition processes, the morphology of the synthetized material, and on the synthesis parameters such as substrate temperature, O_2/Ar gas flow ratio, and sputter gas pressure [11]. Thus it was shown for DC sputtered films that high sputter gas pressure and low substrate temperature enhanced the EC properties in TiO_2 and $a TiO_2$ films. The EC effect in WO_3 thin films is well understood and widely studied both in the sixfold-coordinated crystalline structure $c WO_3$ and in amorphous $a WO_3$ [3]. In $c WO_3$, the ion-e^- inter/deintercalation sites are located inside the empty spaces due to the octahedral arrangement of atoms. The inserted ion-e^- pairs enter extended states, and reflective OM is described by Drude-like free electron absorption models [12]. In $a WO_3$, the OM upon ion-e^- inter/deintercalation occurs in the visible and near-infrared spectral range and leads to an absorption band centered at ~1.3 eV. It has been pointed out that such a behavior is due to variations in the valence states of W atoms. Thus the optical absorption band is modeled by hopping of small polarons between first-neighbor W sites exhibiting different W^{4+}-W^{5+}-W^{6+} oxidation states [13,14]. It was found that aWO$_3$ displays superior OM and EC properties and inherent stability under extended inter/deintercalation of ions. Enhancements of the EC properties and durability have been displayed in DC sputter-deposited W-Ti oxide films [15].

Below we report on the OM and EC properties in stoichiometric and oxygen-deficient $a\mathrm{WO}_{3-x}$ and $a\mathrm{TiO}_{2-x}$ thin films under Li^+-ion-e^- inter/deintercalation. We investigate the substoichiometry dependence on the optical absorption, the EC properties, and the OM for DC sputtered oxygen-deficient films in as-deposited $a\mathrm{WO}_{3-x}$ and $a\mathrm{TiO}_{2-x}$, and in fully Li^+-ion-e^- intercalated $a\mathrm{Li}_y\mathrm{WO}_{3-x}$ and $a\mathrm{Li}_y\mathrm{TiO}_{2-x}$ states by in-situ optical transmittance.

2. Experimental

A number of $a\mathrm{WO}_{3-x}$ and $a\mathrm{TiO}_{2-x}$ films were prepared by reactive DC magnetron sputtering from 99.95%-pure metallic Ti and W targets, using a Balzers UTT-400 unit, in O_2/Ar reactive plasma of 99.99% purity at $\sim 1.5\times 10^{-2}$ mbar. Stoichiometric and oxygen-deficient films were deposited onto unheated glass substrates pre-coated with $\mathrm{In_2O_3}$:Sn. Depositions were done at the same pressure, power and Ar-flow-ratio but at different O_2-flow-ratios. X-ray diffraction analysis indicated amorphous structures. Atomic compositions and density were determined by use of 2MeV ^{4}He-ions at ~170° by Rutherford Backscattering Spectroscopy. The $a\mathrm{WO}_{3-x}$ and $a\mathrm{TiO}_{2-x}$ films had thicknesses of ~300 nm, as verified by surface profilometry measurements (Bruker DektakXT step profilometer). Surface nanotopography and morphological features were investigated by Atomic Force Microscopy (AFM) employing a PSIA XE150 SPM/AFM instrument having an etched cantilever with a tip radius of ~10 nm and ~35° apex angle. Scans and evaluation of root-mean-square roughness were carried out over areas of ~1000×1000 nm.

Li^+-ion-e^- pairs inter/deintercalation was carried out in a glove box containing Ar gas (H_2O<4 ppm). Electrochemical measurements were performed in a three-electrode cell with the $a\mathrm{WO}_{3-x}$ or $a\mathrm{TiO}_{2-x}$ films as working electrode and metallic Li^+-foils as counter and reference electrodes immersed in 1M $\mathrm{LiClO_4}$-$\mathrm{C_4H_6O_3}$ electrolyte. Cyclic voltammetry (CV) measurements were conducted using an ECO Chemie Autolab/GPES Electrochemical Interface with voltage sweep rate of 10 mV/s in the voltage range 2.0-4.0 V vs. Li. The measurements were performed for 30 cycles in order to achieve reversibility of the inter/deintercalation process. Optical properties of $a\mathrm{WO}_{3-x}$, $a\mathrm{TiO}_{2-x}$, $a\mathrm{Li}_y\mathrm{WO}_{3-x}$ and $a\mathrm{Li}_y\mathrm{TiO}_{2-x}$ films during Li^+-ion-e^- inter/deintercalation were measured as in-situ normal incidence transmittance $T(\lambda)$ spectra by using an Ocean Optics instrument.

3. Results and Discussion

Figure **1** shows a characteristic AFM image of an as-deposited $a\mathrm{WO}_{3-x}$ thin film. The image exhibits a smooth surface characterized by a flat topography with granular traits and porous structure. Variations in particle sizes and granular/porous features were slightly dependent on the O_2/Ar ratio. Figure **2** reports $T(\lambda)$ for $a\mathrm{WO}_{3.00}$ films obtained at O_2/Ar=44%, $a\mathrm{WO}_{2.93}$ at O_2/Ar=36%, and $a\mathrm{WO}_{2.84}$ at O_2/Ar=16%; data are shown at $380\leq\lambda\leq 800$ nm for as-deposited, Li^+-ion-e^- intercalated (dark-blue-colored), and bleached films. The stoichiometric $a\mathrm{WO}_{3.00}$ films are transparent and exhibit a high value of $T(\lambda)$ of around ~94%. Pronounced optical absorption can be observed as the oxygen deficiency is increased. As-deposited $a\mathrm{WO}_{2.93}$ and $a\mathrm{WO}_{2.84}$ films display a clear-blue color, and $T(\lambda)$ is reduced to ~86% and ~58% at λ=500 nm, respectively. Upon electrochemical Li^+-ion-e^- intercalation, all films attain a dark-blue color. The transmittance is decreased in the whole spectral range and displays a strong and broad optical absorption. In particular, the $a\mathrm{WO}_{3.00}$ and $a\mathrm{WO}_{2.93}$ thin films display good coloration efficiency and good reversibility from dark-blue-colored to bleached states, which confirms their suitable EC properties. However, the more oxygen-deficient films of $a\mathrm{WO}_{2.84}$ did not exhibit as good EC performance and OM behavior since the specimen did not bleach efficiently upon charge deintercalation. The variation in $T(\lambda)$, between dark-blue-colored and bleached states is highest for $a\mathrm{WO}_{3.00}$ (around ~36 to 86% at λ=550 nm) but is lower for the substoichiometric films of $a\mathrm{WO}_{2.93}$ (~26 to 77% at λ=550 nm) and $a\mathrm{WO}_{2.84}$ (~8 to 48% at λ=550 nm). Charge capacity obtained from CV data was found to be in the range ~16.8 to 18.2 mC/cm^2 (cf. figure **3**) and was slightly influenced by the surface morphology of the films deposited at different O_2/Ar ratios. The fraction of intercalated Li^+-ions per formula unit of W-ions was found to be somewhat different among films of $Li_{0.72}WO_{3.00}$, $Li_{0.66}WO_{2.93}$ and $Li_{0.63}WO_{2.84}$; this could be due to small density variations which influence the ion diffusion processes.

A similar analysis was carried out for as-deposited $a\mathrm{TiO}_{2-x}$ thin films. Surface morphology reveals analogous smooth features with flat topography and granular-like porous structure, as shown in the AFM image in figure **4**. Stoichiometric $a\mathrm{TiO}_{2.00}$ thin films, obtained at a flow ratio for O_2/Ar of 4%, exhibit a $T(\lambda)$ of ~90%, i.e., a transparent state. The oxygen-deficient $a\mathrm{TiO}_{1.87}$

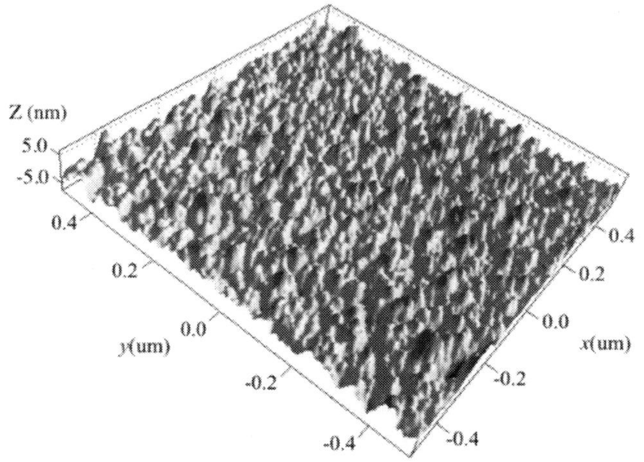

Figure 1. Characteristic AFM image of an as-deposited $a\mathrm{WO}_{3-x}$ thin film.

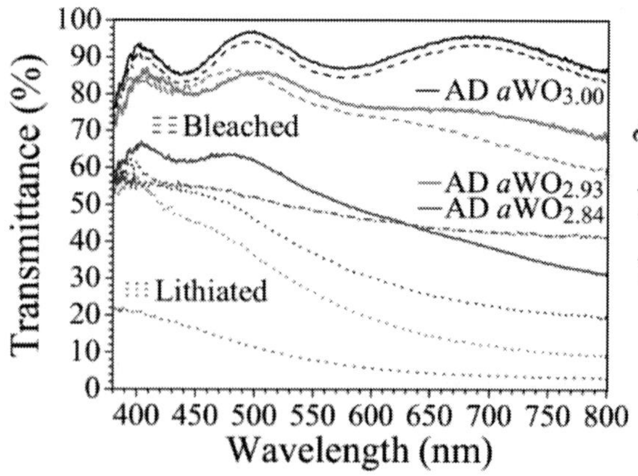

Figure 2. In-situ normal $T(\lambda)$ spectra for $a\mathrm{WO}_{3.00}$, $a\mathrm{WO}_{2.93}$, and $a\mathrm{WO}_{2.84}$ films.

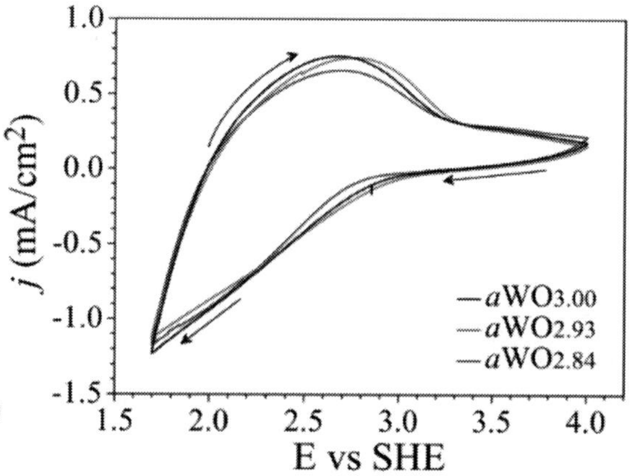

Figure 3. CV curves (at the 10th cycle) for $a\mathrm{WO}_{3.00}$, $a\mathrm{WO}_{2.93}$, and $a\mathrm{WO}_{2.84}$ films.

films, obtained at an O_2/Ar flow ratio of 1.5%, exhibit a transmittance of ~78% at λ=550 nm and displayed a clear-blue colored state with slight optical absorption. Upon electrochemical Li^+-ion-e^- pairs inter/deintercalation, it is found that the aTiO$_{2.00}$ film displays suitable EC properties with good OM and reversibility. Variations in $T(\lambda)$ spectra from dark-blue-colored to bleached state are observed in the range ~40 to 90% at λ=550 nm. For the oxygen-deficient aTiO$_{1.87}$ films, EC properties, OM and reversibility are also present, but the specimen did not bleach efficiently upon charge deintercalation. Optical transmittance variations are displayed in the range ~16 to 67% at λ=550 nm, as indicated in figure **5**. CV data yielded charge capacity in the range ~28.3 to 32.6 mC/cm^2, and the fraction of intercalated Li^+-ions per formula unit of Ti-ions was $Li_{0.65}TiO_{2.00}$ and $Li_{0.60}TiO_{1.87}$ (cf. figure **6**); these values could be slightly influenced by the density increases in films at low O_2/Ar deposition ratios.

Our results suggest that aWO$_{3-x}$ and aTiO$_{2-x}$ thin films display suitable OM and EC properties when deposited at high O_2/Ar ratios. Increasing oxygen content results in nanoporous-like structures in as-deposited films in which pore orientation, homogeneity and

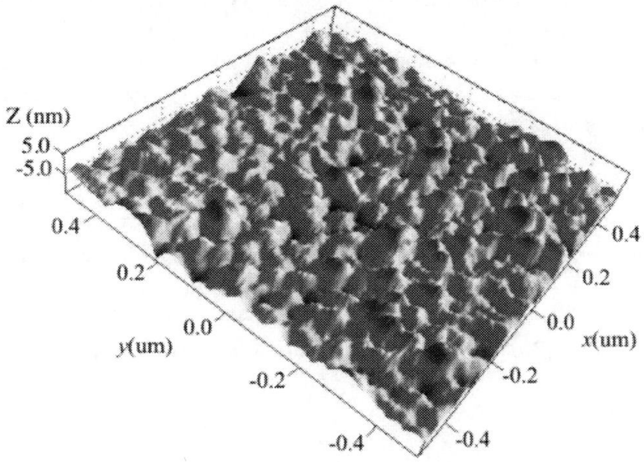

Figure 4. Characteristic AFM image of an as-deposited aTiO$_{2-x}$ thin film.

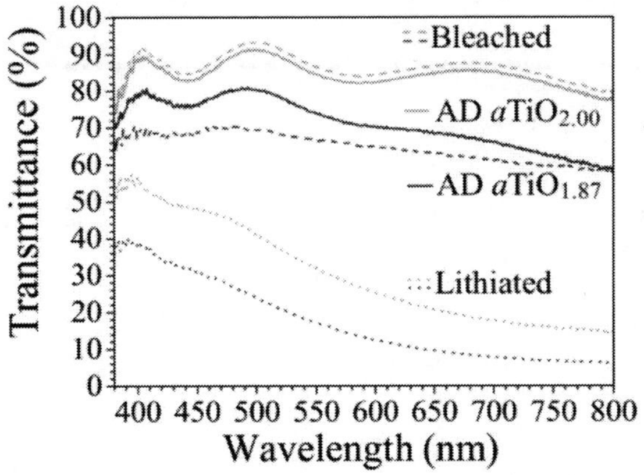

Figure 5. In-situ normal $T(\lambda)$ spectra for aTiO$_{2.00}$ and aTiO$_{1.87}$ thin films.

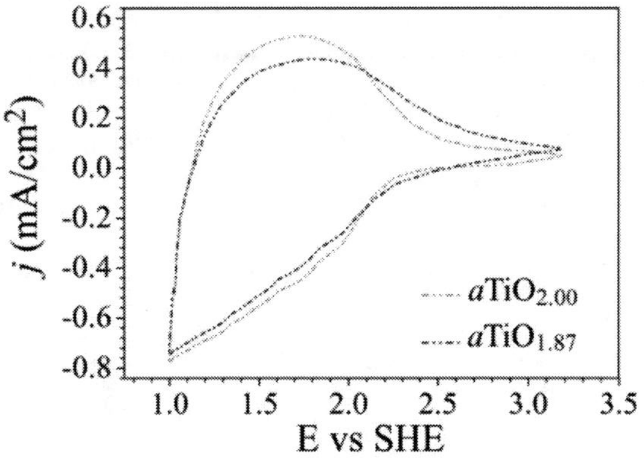

Figure 6. CV curves (at the 10th cycle) for aTiO$_{2.00}$ and aTiO$_{1.87}$ thin films.

symmetry enable ion diffusion across the amorphous network of the films, hence improving their EC behavior. Increasing oxygen deficiency results in thin films exhibiting clear-blue or even dark-gray-colored appearances, which are not suitable for EC applications since the capability for Li^+-ion-e^- inter/deintercalation decreases considerably and the films display incomplete bleaching upon deintercalation of charge. The Li^+-ion diffusion and the fraction of intercalated ions can be strongly influenced by increases in density and variation in porosity. Since EC efficiency depends on the morphology of the amorphous network in the as-deposited $a\mathrm{WO}_{3-x}$ and $a\mathrm{TiO}_{2-x}$ thin films, it is of fundamental importance to ensure low deposition temperature, high O_2/Ar gas flow ratio and suitable deposition pressure in order to achieve the microstructural properties favoring the good optical and EC performance of these films.

Conclusions

We have carried out a study of EC properties in DC magnetron sputtered stoichiometric and oxygen-deficient $a\mathrm{WO}_{3-x}$ and $a\mathrm{TiO}_{2-x}$ thin films under Li^+-ion-e^- inter/deintercalation. Results show good EC behavior, which is strongly dependent on the oxygen content during the deposition process and slightly dependent on the surface morphology. We found good optical modulation and charge capacity upon Li^+-ion-e^- inter/deintercalation for almost stoichiometric films. However, strongly oxygen deficient $a\mathrm{WO}_{3-x}$ and $a\mathrm{TiO}_{2-x}$ films do not display good EC properties but show strong optical absorption also in the bleached state. Our results for stoichiometric films are comparable with those in previous studies carried out on amorphous EC thin films prepared by various deposition methods.

Acknowledgments

This work has been supported by the Swedish Research Council and by the European Research Council under the European Community's Seventh Framework Program (FP7/2007-2013)/ERC Grant Agreement No. 267234 (GRINDOOR). The paper was presented at the INERA Workshop "Transition Metal Oxide Thin Films: Functional Layers in 'Smart Windows' and Water Splitting Devices", September 4-6th, 2014, Varna, Bulgaria. The Workshop is part of the Program of the INERA REGPOT Project at the Institute of Solid State Physics, Bulgarian Academy of Sciences.

References

[1] C.G Granqvist 1995 *Handbook of Inorganic Electrochromic Materials* (Amsterdam: The Netherlands, Elsevier)
[2] P.M.S. Monk, R.J, Mortimer, and D.R. Rosseinky 1995 *Electrochromism: Fundamentals and Applications* (New York: VCH)
[3] C.G. Granqvist 2014 *Thin Solid Films.* **564** 1
[4] H. Fan, A. Wright, J. Gabaldon, A. Rodriguez, C. Brinker, Y. B. Jiang 2006 *Adv. Funct. Mater.* **16** 981
[5] T. Seike, J. Nagai 1991 *Sol. Energy Mater. Sol. Cells.* **22** 107
[6] M. Kitao, Y. Oshima, K. Urabe 1997 *Japan. J. Appl. Phys.* **36** 4423
[7] N.N. Dinh, N.Th.T.Oanh, P.D. Long, M.C. Bernard, A. Hugot-LeGoff 2003 *Thin Solid Films.* **423** 70
[8] C.M. Wang, S.Y. Lin, Y.C. Chen 2008 *J. Phys. Chem. Solids.* **69** 451
[9] K. Lee, D. Kim, S. Berger, R. Kirchgeorg, P. Schmuki 2012 *J. Mater. Chem.* **22** 9821
[10] N.N. Dinh, D.H. Ninh, T.T. Thao, V.V. Truong 2012 *J. Nanomater.* **781236** 1
[11] I. Sorar, E. Pehlivan, G.A. Niklasson, C.G. Granqvist 2013 *Sol. Energy Mater. Sol. Cells.* **115** 172
[12] J.S.E.M. Svensson, C.G. Granqvist 1984 *Appl. Phys. Lett.* **45** 828
[13] L. Berggren, A. Azens, G.A. Niklasson 2001 *J. Appl. Phys.* **90** 1860
[14] M. Merz, J. Eisenmenger, B. Heinz, P. Ziemann 2002 *Phys. Rev. B.* **66** 184102
[15] M.A. Arvizu, C.A. Triana, B.I. Stefanov, C.G. Granqvist, and G.A. Niklasson 2014 *Sol. Energy Mater. Sol. Cells.* **125** 184

Durability of VO_2-based thin films at elevated temperature: Towards thermochromic fenestration

Yu-Xia Ji, Gunnar A Niklasson and Claes G Granqvist

Department of Engineering Sciences, The Ångström Laboratory, Uppsala University, P.O. Box 534, SE-75121 Uppsala, Sweden

Corresponding author: yuxia.ji@angstrom.uu.se

Abstract. An explorative study was performed on sputter-deposited thermochromic VO_2 films with top coatings of Al oxide and Al nitride. The films were exposed to dry air at a high temperature. Bare 80-nm-thick VO_2 films rapidly converted to non-thermochromic V_2O_5 under the chosen conditions. Al oxide top coatings protected the underlying VO_2 films and, expectedly, increased film thickness yielded improved protection. Specifically, it was found that a 30-nm-thick sputter-deposited Al oxide top coating delayed the oxidation by more than one day upon heating at 300°C. The results demonstrate the importance of protective layers in thermochromic windows for practical application.

1. Introduction

Vanadium dioxide has a reversible metal–insulator transition at a "critical" temperature τ_c of about 68 °C [1]. Below this temperature, at $\tau < \tau_c$, VO_2 is monoclinic, semiconducting and infrared-transparent whereas it is tetragonal, metallic-like and infrared-reflecting at $\tau > \tau_c$. These properties imply that thin films of VO_2 exhibit thermochromism and can be of interest for windows and glass facades in order to control the inflow of visible light and solar energy into energy efficient buildings.

Thermochromic (TC) thin films of VO_2 have been discussed as adaptive coatings for glazing applications for several decades [2–5]. However, the progress towards practical applications has been disappointingly slow because of performance deficiencies such as a too high value of τ_c, too large luminous absorptance and too small modulation of the solar energy throughput [6]. However, recently there has been substantial progress towards the alleviation of all of these obstacles [7–9]. For practical implementation, the TC VO_2-based films must maintain their desirable properties for long times. This issue requires particular attention since the structure of VO_2 is not thermodynamically stable, and the material may be further oxidized and form non-TC V_2O_5 at high temperature and under humid conditions [10–12].

The most accurate method to test the durability of the coating is exposing it to the natural working conditions, but it is clearly very time-consuming to get satisfactory results. Instead, tests at high temperatures can be utilized to bring the aging time scale into a practical regime. Hence, accelerated aging tests are inexpensive and timesaving. In the present work we report on explorative studies of sputter-deposited VO_2 films subjected to temperatures up to 300 °C in dry air and demonstrate that thin sputter-deposited top coatings of Al oxide can provide good protection for TC VO_2. The present paper reports data that are complementary to those in an earlier study of ours [12].

2. Experimental

2.1 Film preparation

Thin films of vanadium dioxide, aluminum oxide and aluminum nitride were deposited from 5-cm-diameter vanadium (99.95% purity) and aluminum (99.99% purity) targets using reactive DC magnetron sputtering onto 1-mm-thick glass substrates. A detailed description of the film preparation can be found elsewhere [12].

In short, the VO_2 layer was first deposited at a power of 172 W in a mixture of argon and oxygen wherein the oxygen/argon gas flow ratio was kept at a constant value of 0.05 during the deposition process while the process pressure in the growth chamber was held constant at 1.2×10^{-2} mbar. The glass substrates were maintained at 450°C during deposition, as measured by a thermocouple. After deposition, the sample was cooled under vacuum.

The thickness of the deposited VO_2 films was kept constant at 80 nm; it was measured using a Dektak XT stylus profilometer. The deposition rate was 7 nm/minute.

A protective layer of Al_2O_3 was deposited onto the unheated VO_2 film by reactive DC magnetron sputtering from a pure aluminum target using an oxygen/argon gas flow ratio of 0.025 while the working pressure in the growth chamber was set at 4×10^{-2} mbar. The power on the target was 200 W and the deposition rate was 20 nm/minute.

Similarly to the aluminum oxide layer, an aluminum nitride layer was deposited onto the unheated VO_2 film by DC sputtering in a nitrogen/argon gas flow ratio of 0.025. The power on the target was 200W and the deposition rate was 25 nm/minute.

2.2 Accelerated aging test

To investigate the durability of the TC films, the samples were heated in a horizontal quartz-tube furnace. The temperature of the furnace center was kept at 300 °C under the flow of dry air at a rate of 100 sccm. After heating, the samples were cooled to room temperature in nitrogen atmosphere.

After each accelerated aging treatment, the optical transmission of the as-prepared and aged samples was recorded by a Perkin-Elmer Lambda 900 spectrophotometer. The transmittance was measured in the wavelength interval $300 < \lambda < 2500$ nm at room temperature and at an elevated temperature of 100 °C.

3. Results

Figure 1 shows spectral transmittance of a VO_2 film in as-deposited state and after treatment for one hour in air at 300 °C. Panels (a) and (b) refer to $\tau < \tau_c$ and $\tau > \tau_c$, respectively. The as-deposited film shows clear thermochromism and exhibits much higher infrared transmittance at room temperature than at 100°C.

Heat-treated films display very different optical properties, and no trace of thermochromism can be seen. The latter data are consistent with those expected for V_2O_5 [13], and it is evident that the vanadium-based film is fully oxidized.

Figure 2 reports analogous results of spectral transmittance for VO_2 films coated with a 10-nm-thick layer of Al nitride. These data indicate that the thermochromism has disappeared after 1 h.

This might be due to strain induced during the heat treatment process and thus the VO_2 film was stabilized in the metallic state. Therefore the Al nitride coating could not act as viable anti-oxidation layer for the thermochromic VO_2 films.

Journal of Physics: Conference Series **559** (2014) 012005 doi:10.1088/1742-6596/559/1/012005

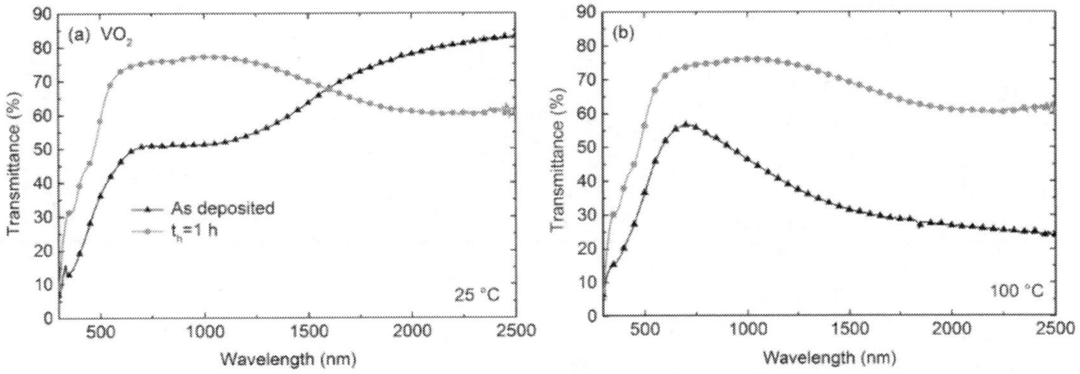

Figure 1. Spectral transmittance for an 80-nm-thick VO_2 film before and after heating at 300°C for one hour. Data were taken at $\tau < \tau_c$ (panel a) and $\tau > \tau_c$ (panel b). The results were reported also in earlier work of ours [12].

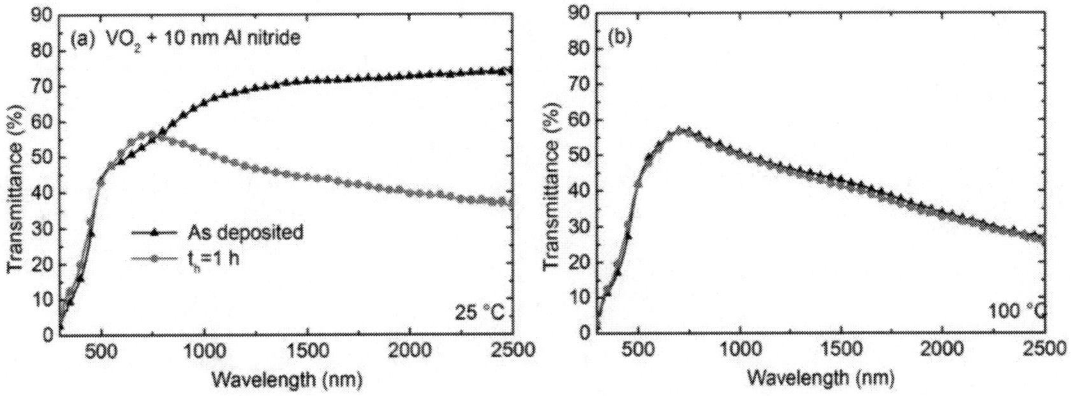

Figure 2. Spectral transmittance for an 80-nm-thick VO_2 film, coated with 10 nm of Al nitride, before and after heating at 300 °C for one hour. Data were taken at $\tau < \tau_c$ (panel a) and $\tau > \tau_c$ (panel b).

Figure 3 shows data for VO_2 films coated with Al oxide of the thicknesses 10 nm (panels a and b) and 30 nm (panels c and d) after heating in air at 300°C for the shown durations t_h.

Clearly the Al oxide prevents oxidation of the underlying VO_2, and the optical data are almost unchanged for the thicker Al oxide layer whereas the sample with the thinner top coating started to undergo some changes.

A minor increase in the spectral transmittance for t_h =30 h can be reconciled with a conversion of a slight amount of VO_2 to V_2O_5.

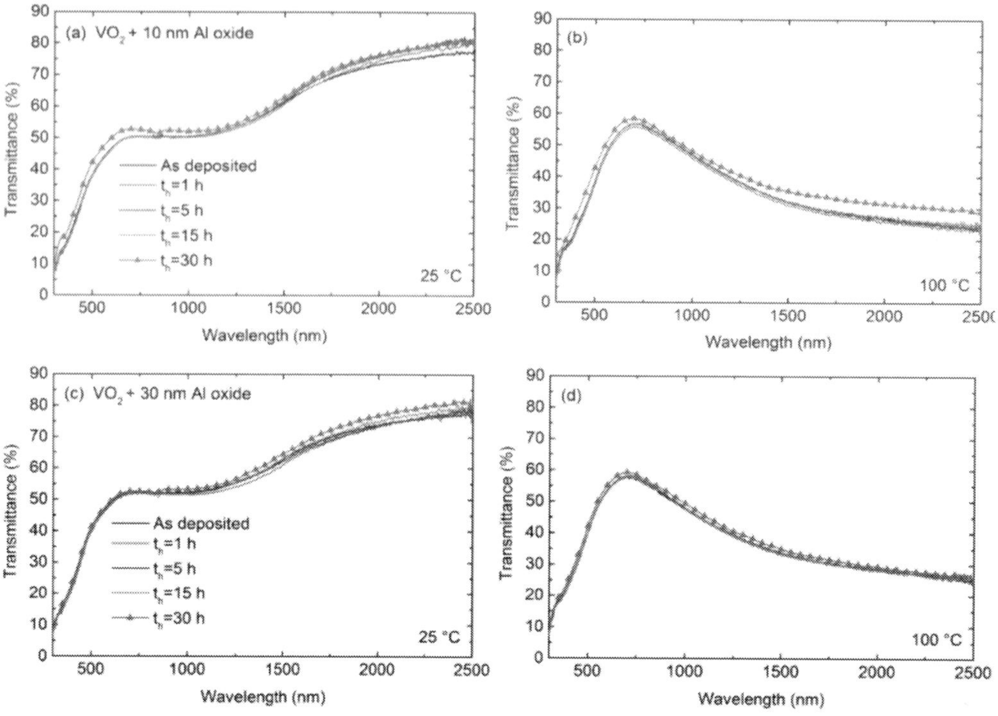

Figure 3. Spectral transmittance for 80-nm-thick VO_2 films, coated with 10 nm (panels a and b) and 30 nm of Al oxide (panels c and d), in as-deposited state and after heating at 300 °C in air for the shown durations t_h. Data were taken at $\tau < \tau_c$ (panels a and c) and $\tau > \tau_c$ (panels b and d). The results were reported also in earlier work of ours [12].

Further studies were conducted also for thicker Al oxide films, and figure 4 reports on a VO_2 film coated with 100 nm of Al oxide and heat treated in air at 300°C for 168 h. It is clear that essentially all of the thermochromism remains after this time, *i.e.*, after one week.

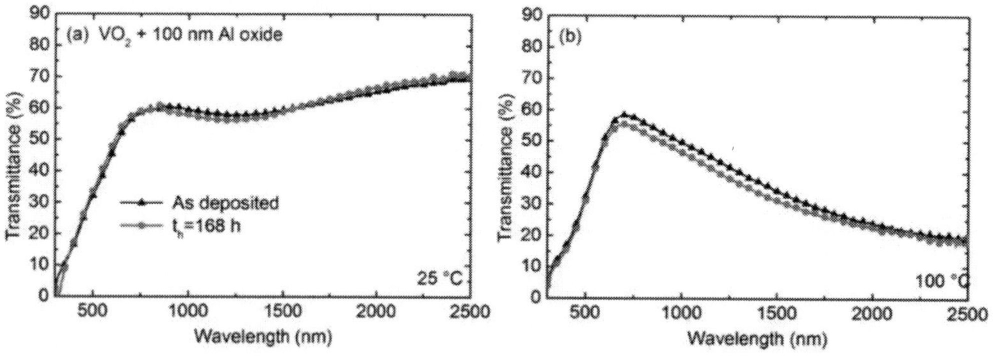

Figure 4. Spectral transmittance for an 80-nm-thick VO_2 film, coated with 100 nm of Al oxide, in as-deposited state and after heating at 300 °C in air for the shown durations t_h. Data were taken at $\tau < \tau_c$ (panel a) and $\tau > \tau_c$ (panel b).

4. Remarks and conclusion

A bare VO_2 film could not maintain its thermochromism under accelerated aging in dry air at 300 °C. The oxidation mechanism might be that tetragonal VO_2 (β-phase, space group $P4_2/mnm$, at $\tau > 68$ °C) transforms progressively into monoclinic V_6O_{13} ($C2/m$ at $\tau > -124$ °C) and V_3O_7 ($C2/c$), until ultimately reaching orthorhombic V_2O_5 (*Pmnm*), as inferred from the equilibrium phase diagram of the oxygen–vanadium system [10,11].

A VO_2 film coated with 10 nm of Al oxide kept its thermochromism fairly well for heat treatment up to 30 h at 300 °C, and a thicker Al oxide coating of 30 nm yielded enhanced oxidation protection, as expected. A 10-nm-thick top layer of Al nitride yielded a radically different performance and could not offer a similar oxidation shielding under the chosen conditions.

The Al oxide top coating on thermochromic VO_2 films is not only working as an oxidation-preventing layer but can be used for anti-reflection purposes to enhance the luminous transmittance and thereby provide properties of interest for practical thermochromic fenestration [9].

Acknowledgements

Financial support was received from the European Research Council under the European Community's Seventh Framework Program (FP7/2007–2013)/ERC Grant Agreement No. 267234 (GRINDOOR).

References

[1] Morin F J 1959 *Phys. Rev. Lett.* **3** 34–36
[2] Goodenough JB 1971 *J. Solid State Chem.* **3** 490–500
[3] Jorgenson GV and Lee J C 1986 *Sol. Energy Mater.* **14** 205–214
[4] Babulanam SM *et al* 1987 *Sol. Energy Mater.* **16** 347–363
[5] Saeli M *et al* 2010 *Sol. Energy Mater. Sol. Cells* **94** 141–151
[6] Gao Y *et al* 2012 *Nano Energy* **1** 221–246
[7] Li SY *et al* 2012 *Thin Solid Films* **520** 3823–3828
[8] Zhou J *et al* 2013 *Phys. Chem. Chem. Phys.* **15** 7505–7511
[9] Li SY *et al* 2014 *J. Appl. Phys.* **115** 053513/1–053513/10
[10] Smith JF (editor) 1989 *ASM International Monograph Series on Alloy Phase Diagrams*, (Metals Park, OH, USA)
[11] Wriedt HA 1989 *Bull. Alloy Phase Diagrams* **10** 271–277
[12] Ji YX *et al* 2014 *Thin Solid Films* **562** 568–573
[13] Talledo A and Granqvist CG 1995 *J. Appl. Phys.* **77** 4655–4666

Electrochromic performance of Ni oxide thin films intercalated with Li^+ ions

Rui-Tao Wen[1], **Gunnar A Niklasson and Claes G Granqvist**
Department of Engineering Sciences, The Ångström Laboratory, Uppsala University
P. O. Box 534, SE-75121 Uppsala, Sweden

E-mail: Ruitao.Wen@angstrom.uu.se

Abstract. Porous Ni oxide thin films were deposited on unheated ITO/glass substrates by sputtering in argon–oxygen. The as-deposited thin films have a cubic NiO structure and still exhibit such a structure after 10,000 electrochemical cycles in 1 M $LiClO_4$ in propylene carbonate in the range between 2.0 and 4.1 V *vs* Li/Li^+. Electrochromic performance showed a rapid drop of charge density over the first hundreds of cycles and subsequently a very slow decrease. The charge density was 87% of the initial one after 1,000 cycles and 82% after 10,000 cycles, indicating an extremely slow decay after 1,000 cycles. Optical modulation was also slightly decreased after 10,000 cycles, which is due to the drop of charge density.

1. Introduction

Energy consumption in buildings is a very urgent issue, since the fraction of the world's primary energy that is used in buildings amounts to as much as 30–40% [1]. Traditional windows are often regarded as weak components for efficient energy control in buildings [2]. However, switchable windows, also known as smart windows [3], are very promising for alleviating the energy consumption effectively in buildings [4, 5]. Specifically, "smart" windows based on electrochromism can regulate the amount of sunlight that can pass through the window when an external, small potential is applied. Therefore, electrochromic (EC) thin films have attracted intensive interest and have been widely studied in recent years. The most important requirements for smart windows are large optical modulation in the solar spectrum and long-term durability.

Ni oxide is an anodic EC material and has great potential in smart windows. Most previous research work was performed on Ni oxide thin films in aqueous electrolyte systems (mostly potassium hydroxide, KOH) [3], and it was found that Ni oxide has large optical modulation [6], high coloration efficiency [7], good durability [8], and fast switching time [7]. In some cases, Ni oxide films also served as counter electrode to WO_3 [9] in EC windows in order to get increased optical modulation, since tungsten oxide darkens under charge insertion while Ni oxide darkens under charge extraction [10]. In addition, the color appearance is improved because the brown color of Ni oxide is complementary to the blue color of tungsten oxide, both together giving a neutral (gray) appearance in the colored state. However, tungsten oxide and Ni oxide are chemically incompatible when they are working in an acidic or alkaline aqueous electrolyte environment [11]. Therefore, Ni oxide needs to be compatible with tungsten oxide in a non-aqueous electrolyte system, where $LiClO_4$ in propylene

[1] Corresponding author: Ruitao.Wen@angstrom.uu.se

carbonate (Li–PC) is mostly used in WO_3-based EC devices. So far, a number of investigations have been published on EC performance of Ni oxide in non-aqueous electrolyte [12–14]. However, long-term durability of Ni oxide in Li^+-based electrolyte has seldom been reported. In fact, the long-term stability is one of the most critical issues for commercial use of Ni-oxide-containing smart windows.

In this report, Ni oxide thin films of thickness ~300 nm were deposited on unheated substrates by reactive dc magnetron sputtering. The working pressure was kept high in order to obtain porous thin films [15]. EC properties were characterized by cyclic voltammetry (CV) and optical measurements.

2. Experimental

Film deposition — The Ni oxide films were deposited by reactive dc magnetron sputtering in a system based on a Balzers UTT 400 unit. The target was a 5-cm-diameter cylindrical plate of metallic nickel (99.95%). Sputtering was performed in an argon–oxygen atmosphere at room temperature. Pre-sputtering took place in argon (99.998%) for 5 minutes, and oxygen (99.998%) was then introduced. During deposition, the O_2/Ar gas-flow ratio was kept at a constant value of 3.5%. The total pressure during sputtering was set to ~30 mTorr, and the power at the target was 200 W. The substrates were glass plates pre-coated with transparent and electrically conducting layers of In_2O_3:Sn (known as ITO) with a sheet resistance of 60 Ω. Carbon substrates were also used to deposit films for Rutherford Backscattering Spectrometry (RBS) measurements. The substrates were rotated during deposition to ensure even films. The film thickness d was measured by surface profilometry using a DektakXT instrument.

Physical techniques — Crystalline structures were investigated by X-ray diffraction (XRD), using a Siemens D5000 diffractometer with CuK_α radiation at a wavelength of 0.154 nm. Structure and phase composition were obtained by comparison with the Joint Committee on Powder Diffraction Standards (JCDPS) data base. Morphology of the films was characterized by scanning electron microscopy (SEM) using a LEO 1550 FEG Gemini instrument with an acceleration voltage of 10 kV. Density and atomic concentrations of the films were determined by RBS at the Uppsala Tandem Laboratory, using 2MeV 4He+ ions back scattered at an angle of 170 degrees. The density was calculated from $(M \cdot N_s)/(n_a \cdot N_A \cdot d)$, where M is molar mass, N_s is areal density of atoms, na is the number of atoms in a molecule, and N_A is Avogadro's constant. The density was found to be ~4.2 g/cm^3, indicating that the films are highly porous compared to bulk NiO (6.8 g/cm^3).

Electrochemical and in situ optical characterization — CV was carried out in a conventional three-electrode electrochemical cell by use of a computer-controlled ECO Autolab/GPES Interface. Ni oxide served as the working electrode in an electrolyte of 1 M Li–PC. Both counter and reference electrodes were Li foils, and the potential scans were performed between 2.0 and 4.1 V *vs* Li/Li^+ with a sweep rate of 50 mV/s. The low potential limit was set to 2.0 V in order to get an easy comparison with pervious research on iridium oxide [16]. Optical transmittance was recorded *in situ* during electrochemical cycling of Ni oxide films in the 380–800 nm wavelength range by using a fiber-optical instrument from Ocean Optics. The electrochemical cell was positioned between a tungsten halogen lamp and the detector, and the 100-%-level was taken as the transmittance before immersion of the sample in the electrolyte.

3. Results and Discussions

XRD patterns of as-deposited Ni oxide and the same film after 10,000 cycles are compared in figure 1. A spectrum of an ITO coated glass is also shown as a reference. The as-deposited Ni oxide film exhibits a cubic structure (*Fm3m*), as indicated by the (111), (200), (220), (311) and (220) lattice planes (black curve). No traces of phases containing hydrogen, or other phases, were detected. Grain size was calculated to be ~40 nm by Scherrer's formula [17] using the (111) diffraction peak. The sample still shows the cubic structure (red curve), and no other phase appears after electrochemical cycling for 10,000 times. By carefully comparing the diffraction peaks before and after cycling, we find that the crystallinity of Ni oxide is hardly altered after cycling, which is indicated by the same diffraction intensity of the (111) peak (figure 1b). This means that the structure is still very stable after long-term cycling.

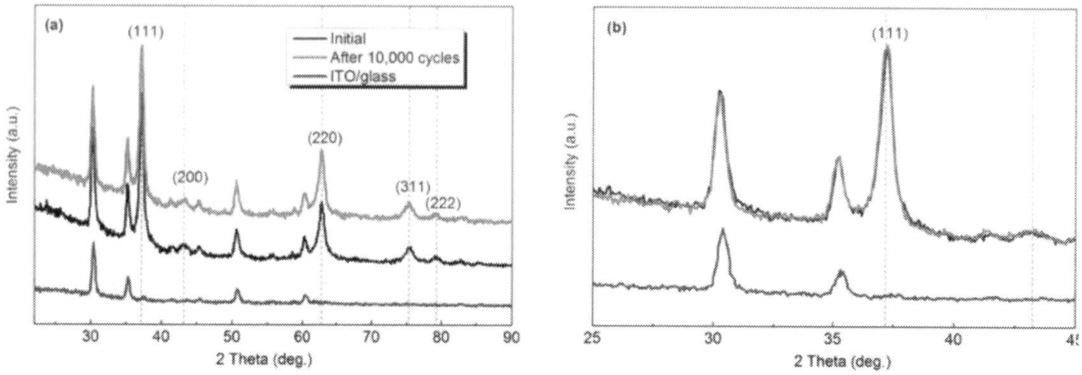

Figure 1. X-ray diffraction of Ni oxide films on ITO coated glass. Panel (a) refers to Ni oxide films before and after 10,000 cycles; peaks are assigned to the cubic phase of NiO (JCPDS No. 04-0835), as indicated by the red dash lines. Panel (b) is an enlargement of panel (a) between 25 and 45°. A diffractogram of the ITO coated glass is shown for comparison..

The surface morphology of the Ni oxide films before and after cycling was characterized by SEM, as illustrated in figure 2. The images show that the films are very homogeneous, and their surfaces have a triangular pyramid-like appearance with a characteristic length of ~40 nm. This result is consistent with the calculated grain size using Scherrer's formula. Clearly, the surface was almost unchanged after electrochemical cycling and still showed triangular pyramids.

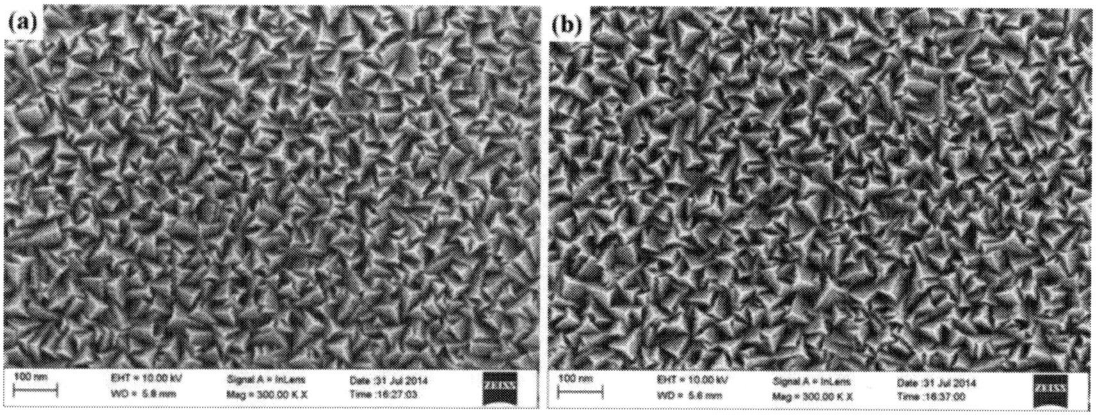

Figure 2. SEM images of Ni oxide thin films (a) before cycling and (b) after 10,000 cycles.

Figure 3 (a) shows CV curves for Ni oxide thin films cycled in 1 M Li–PC. It can be seen that charge insertion and extraction takes place in the whole potential range. There is a persistent shrinking of the CV curve during cycling, implying a decrease of charge insertion and extraction. To be specific, we also plot the inserted and extracted charge density *vs* cycle number in figure 3(b). A rapid drop of charge density can be observed during the first hundreds of cycles; then the charge density drops very slowly upon further cycling, indicating that Ni oxide is stable as an EC layer. *In situ* optical transmittance was recorded between 380 and 800 nm for Ni oxide films, as shown in figure 4. The bleached state was almost unchanged after 10,000 cycles and remained at ~90%. However, the colored-state transmittance was slightly increased. The variation of optical transmittance is in agreement with the decrease of charge density after long-term cycling.

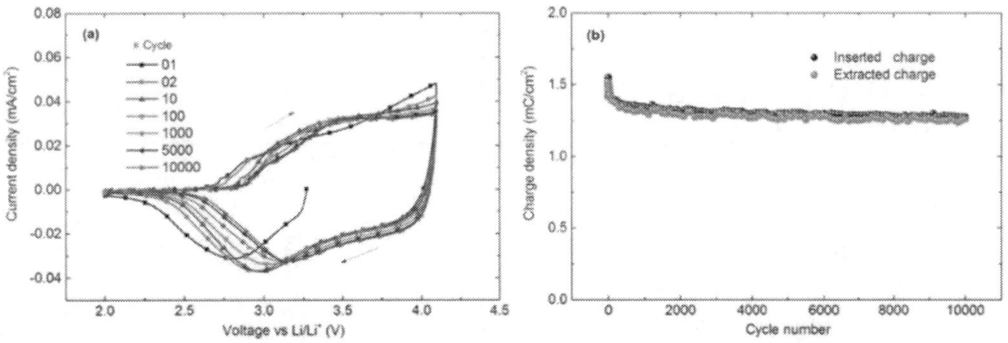

Figure 3. Electrochemical characteristics of the durability of Ni oxide thin films. Panel (a) shows cyclic voltammograms of films in 1 M Li–PC between 2.0 and 4.1 V *vs* Li/Li$^+$; arrows signify voltage sweep direction. Panel (b) illustrates decay of charge density upon cycling.

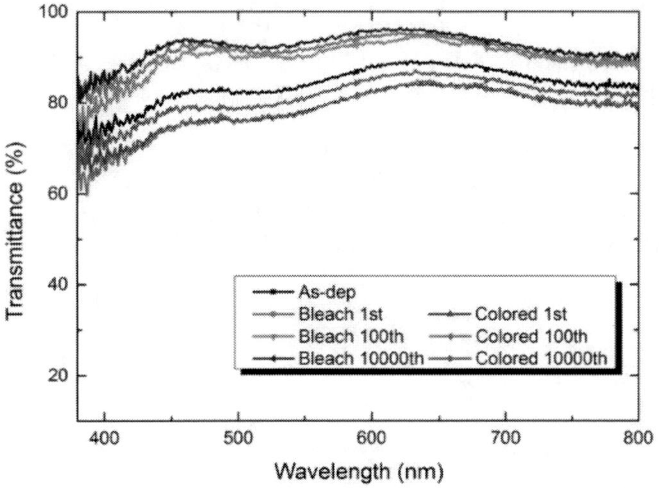

Figure 4. *In situ* optical transmittance of Ni oxide films in 1 M Li–PC before and after 100 and10,000 cycles.

4. Conclusions

In summary, Ni oxide can be used as a stable EC layer in smart windows. XRD shows that the Ni oxide thin films still have a cubic structure after 10,000 electrochemical cycles in Li–PC. A rapid drop of charge density happens only during the first hundreds of cycles, and Ni oxide was found to be very stable upon further electrochemical cycling. A slight decrease of optical modulation was due to the small drop of charge density after long-term cycling.

Acknowledgements

Financial support was received from the Swedish Research Council and from the European Research Council under the European Community's Seventh Framework Program (FP7/2007–2013)/ERC Grant Agreement No. 267234 (GRINDOOR). This paper was presented at INERA Workshop "Transition Metal Oxide Thin Films: Functional Layers in 'Smart Windows' and Water Splitting Devices",

September 4–6th, 2014, Varna, Bulgaria. The Workshop is part of the Program of INERA REGPOT Project of Institute of Solid State Physics, Bulgarian Academy of Sciences.

References

[1] UNEP 2007 *Buildings and Climate Change: Status, Challenges and Opportunities.* (Paris, France)
[2] Baetens R, Jelle B P and Gustavsen A 2010 *Sol. Energy Maer. Sol Cells* **94** 87
[3] Granqvist C G 1995 *Handbook of Inorganic Electrochromic Materials.* (Amsterdam, The Netherlands: Elsevier)
[4] Smith G B and Granqvist C G 2010 *Green Nanotechnology: Solutions for Sustainability and Energy in the Built Environment,* (Boca Raton, FL, USA: CRC Press)
[5] Granqvist C G 2014 *Thin Solid Films* **564** 1
[6] Huang H, Tian J, Zhang W K, Gan Y P, Tao X Y, Xia X H and Tu J P 2011 *Electrochim. Acta* **56** 4281
[7] Yuan Y F, Xia X H, Wu J B, Chen Y B, Yang J L and Guo S Y 2011 *Electrochim. Acta* **56** 1208
[8] Estrada W, Andersson A M and Granqvist C G 1988 *J. Appl. Phys.* **64** 3578
[9] Granqvist C G 2006 *Nature Mater.* **5** 89
[10] Niklasson G A and Granqvist C G 2007 *J. Mater. Chem.* **17** 127
[11] Green S, Backholm J, Georén P, Granqvist C G and Niklasson G A 2009 *Sol. Energy Mater. Sol. Cells* **93** 2050
[12] Passerini S and Scrosati B 1992 *Solid State Ionics* **53–56** 520
[13] Passerini S, Scrosati B and Gorenstein A 1990 *J. Electrochem. Soc.* **137** 3297
[14] Kubo T, Nishikitani Y, Sawai Y, Iwanaga H, Sato Y and Shigesato Y 2009 *J. Electrochem. Soc.* **156** H629
[15] Thornton J A 1977 *Ann. Rev. Mater. Sci.* **7** 239
[16] Wen R-T, Niklasson G A and Granqvist C G 2014 *Sol. Energy Mater. Sol. Cells* **120** 151
[17] Cullity B D and Stock S R 2001 *Elements of X-ray Diffraction, third ed.* (*Prentice-Hall, Upper Saddle River, NJ, USA*)

Journal of Physics: Conference Series **559** (2014) 012007 doi:10.1088/1742-6596/559/1/012007

Fabrication of photonic opal structures on different support materials by convective evaporation

Delphine M Lebrun, Gunnar A Niklasson and Lars Österlund

Department of Engineering Sciences, The Ångström Laboratory, Uppsala University, P.O. Box 534, SE-75121 Uppsala, Sweden

E-mail: dele@angstrom.uu.se

Abstract. Photonic polystyrene (PS) opals with face-centered cubic structure were fabricated by convective evaporation. The influences of substrate and its physical properties, as well as deposition conditions were investigated. It is shown that the surface roughness must be less than about 30% of the bead diameter to form well-ordered opals, rendering substrates such as glass, ITO and quartz superior to thick Kapton, ordinary tape, thermal release semiconductor tape, plastic and hydrophilic plastic HHNW W (Kemafoil). Periodic stripe-like structures were found to form perpendicular to the growth direction defined by the receding meniscus of the solution front when the PS concentration is lower than 1.0 w/v%. Finally, we present the principles and results of using soft sacrificial layer deposited on quartz substrates to fabricate free standing inverse opal structures.

1. Introduction

Artificial opals are self-assembled face-centered cubic (fcc) structures of spherically shaped beads, which show interesting applications as photonic band gap materials [1]. Inverse opals are photonic crystals consisting of fcc packed voids embedded in a high refractive index material matrix. [2-4]. such structures have been used to enhance the efficiency of photocatalytic materials [5-7], and motivates further studies to improve the deposition process and the quality of the template opals [8-12].

Convective evaporation is a common method to make artificial opals [13; 14], and has been used to deposit square centimeters of polystyrene or silica beads. Polystyrene (PS) spheres are readily available in a wide range of sizes and can be deposited on a wide range of substrates. There is however several problems encountered during the deposition process to achieve well-ordered structures[15-17]. Even if the thickness can be controlled relatively well by the oven temperature and humidity as well as the solution concentration [14; 18; 19], it is not straightforward to obtain high quality opals. Due to the strainexerted on the PS beads as the liquid is evaporated from the wet structure, opals tend to crack during the drying process and therefore form millimeter-sized domains [20].

Until now only the hydrophilicity of the substrate has been identified to be a necessary criterion for successful opal ordering [19; 21-23], but in this paper we show that surface roughness is also important. Further, we demonstrate the influence of PS concentration on the film morphology and show that too low concentration leads to strip-like structures with a periodicity of hundreds of micrometers. Finally, we show results of PS self-assembly of opal structures on soft sacrificial layers deposited on quartz substrates, which enables fabrication of free-standing opal structures which can reduce stress in the film upon drying thus facilitating larger domains of well-ordered structures.

2. Experimental

2.1 Materials

All chemicals were used as purchased. All bead suspensions were sonicated in an ultra-sonic bath for at least 30 minutes before use. Quartz (Ted Pella), glass (ThermoScientific), ITO (PGO), Kapton (GoodFellow), hydrophilic plasticHHNW W (Kemafoil), Scotch tape (3M), and Revalpha thermal release tape (Nitto Americas) were used as substrates. Water based polystyrene (PS) suspensions were purchased from Polysciences and ThermoScientific. Ethanol 99% and Decon90 (Decon Laboratories Limited) were used as received. Decon90 contains potassium hydroxide and was used as detergent to clean substrates prior to experiments.

2.2 Sample preparation and characterization

Convective evaporation was employed to deposit polystyrene beads on the different substrates. PS solutions (typically 20 mL) were prepared by dilution in deionized water and subsequent ultra-sonication. The concentrations are presented in units of w/v% defined according to equation 1 (taking into account the solution mass of PS):

$$w/v = \frac{1}{\frac{1-W_w}{\rho_w} + \frac{W_w}{\rho_{PS}}}, \tag{1}$$

where W_w is the weight ratio of (dry) polystyrene to latex, ρ_w is the density of water, and ρ_{PS} is the density of polystyrene.

A clean substrate was placed vertically inside the beaker containing the PS solution at a small angle against the wall of the beaker. The beaker was placed in a pre-heated oven (Memmert UFB400) held at 50°C for between 8 and 14 h depending on the total volume of the solution. After deposition the samples were sintered at 85°C for 1 h 30 min in an oven (Nabathern S17) to create a neck at the PS contact area, which increased the connectivity of the voids in the inverted structure.

Profilometry (Bruker Dektak XT) was used to characterize the surface roughness and the averaged deposition thickness of the sample. The force applied to the stylus was set as low as possible (9.8 mN), and the spatial resolution was approximately 0.33 □m/point. From each profilometry trace, the root mean square (rms) surface roughness was extracted. All features above 1 □m were removed from the data, since they derived from substrates dirtiness.

A Perkin Elmer Lambda900 spectrophotometer equipped with an integrating sphere was used for UV-Vis-NIR spectrophotometry measurements in the range between 200 nm and 800 nm. A step size of 1 nm was employed to scan the wavelength range, and a 5 mm diameter aperture was placed in front of the samples.

To estimate the position of the photonic bandgap peak position the Bragg-Snell law was used (see equations 2 and 3):

$$\lambda = 2d_{hkl}(n_{eff}^2 - \sin^2\theta)^{1/2}, \tag{2}$$

In the case that the close-packed (111) planes are parallel to the top interface, we have:

$$d_{111} = D\sqrt{\frac{2}{3}}, \tag{3}$$

where D is the lattice constant of the photonic crystal, d_{hkl} is the distance between bead planes, n_{eff} is the effective refractive index of the whole structure, and θ is the angle of incidence.

3. Results and discussion

3.1 Influence of substrate surface roughness

To be able to fabricate opals, it has been long established that the surface of the substrate should be hydrophilic. We have tried different types of substrates, all cleaned with Decon90, which provide clean hydrophilic surfaces (contact angle close to zero). However, with similar experimental conditions (temperature, humidity and concentration), it was found to be impossible to fabricate well-ordered opals on several substrates used in this study, including thick Kapton, Scotch tape, and Revalpha thermal release tape, ordinary plastic and hydrophilic plastic HHNW W. Even though those substrates all are hydrophilic, none, or only a few samples, exhibited well-ordered opal fcc domains. In contrast, it was possible to deposit opals on quartz, indium tin oxide (ITO), thin Kapton film and glass, which were pre-treated with Decon90 to give them hydrophilic wetting properties. In all successful deposition experiments on these substrates, domains of extended fcc PS structures formed over the entire substrate.

To understand why the depositions failed for some substrates and were successful in other cases, we decided to look into the surface roughness of the substrates. Figure 1 shows the surface roughness (root mean square, rms, value) of the different substrates from profilometry measurements. It can be seen that quartz, ITO, and glass are smoother than the other substrates. From this study it seems that successful PS opal deposition is achieved for hydrophilic surfaces with surface roughness < 30% of bead diameter.

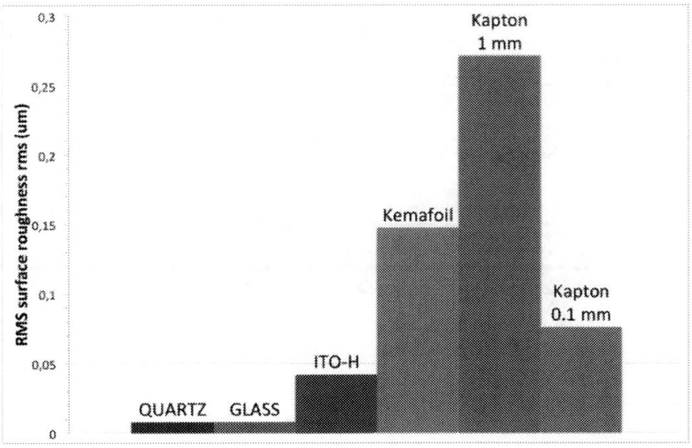

Figure 1. Surface roughness (rms value) for the different substrates used in the present study.

3.2 Formation of polystyrene stripe structures

The deposition of PS beads results in self-assembly of PS beads due to van der Waals interactions whereby PS beads assemble along the solution front. A consequence of this is that it is possible to obtain regularly spaced 15 to 30 layers thick PS ridges of self-assembled PS beads separated by troughs with no beads by using solutions with low concentrations (< 1.0 w/v%) on quartz, as schematically shown in figure 2. These ridges cover the entire length of the solution front across the substrate, and are separated by bare substrate surfaces. On the bare surfaces PS beads have been depleted due to diffusion to complete the fcc structure on adjacent ridges, thus forming stripe-like structures on the substrate with intermittent fcc opal structures and bare substrate surfaces.

Figure 2. Schematic picture of the striped PS structure obtained at low PS concentration. The solution front is parallel to the stripes and it descends as the solution evaporates.

Figure 3 shows typical profilometry traces on quartz. The average peak-to-peak stripe periodicity for different PS diameters and solution concentrations is shown in figure 4. The formation of periodic stripe structures can be explained by the combination of limited supply of beads in the growth process in low concentrated solution with the evaporation mechanisms for self-assembly of beads [24; 25].

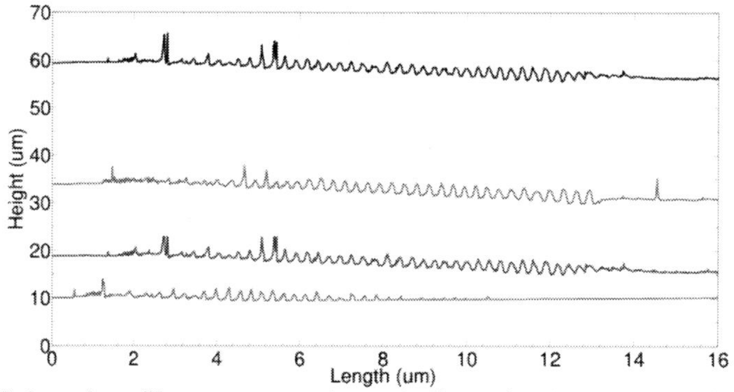

Figure 3. Selected profilometry traces for periodic opal stripe structures formed with 170 nm PS bead diameter and a PS concentration of 0.3 w/v%.

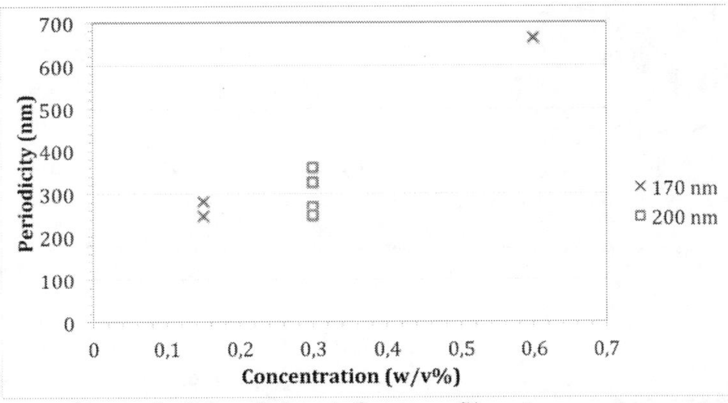

Figure 4. Average stripe periodicity deduced from profilometry measurements for different PS concentrations and PS beads diameters. The stripe periodicity was determined by Fourier transform analysis of profilometry traces such as those shown in figure 3.

Transmission spectra of 170 nm bead diameter PS opals grown with different concentrations of PS (0.15, 0.3, and 0.6 w/v%) are shown in figure 5. A well-ordered opal will have a dip in the transmission spectrum at a certain wavelength due to the presence of a local photonic band gap [26].

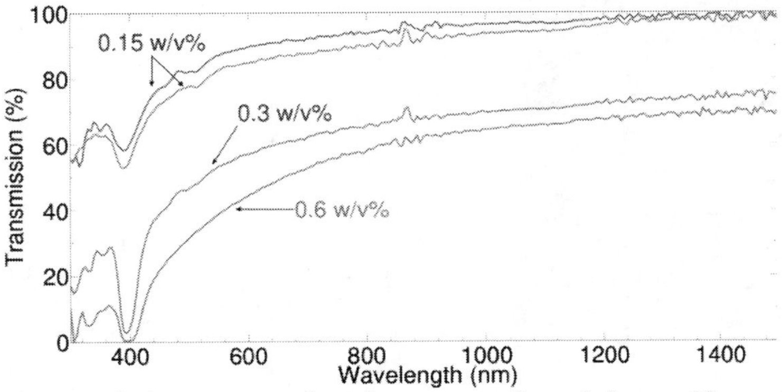

Figure 5. Optical transmission spectra of opal structures formed from self-assembly of 170 nm diameter PS beads at different concentrations.

3.3 Template for fabrication of free-standing inverse opal structures

A possible method to fabricate free-standing opal structures is to grow self-assembled PS opals on a sacrificial layer. With this approach an open opal structure can be developed from the part of the opal which is lifted off from the sacrificial layer (figure 6). With appropriate thickness, we expect that stable, free standing inverse opals can be fabricated.

This approach may potentially alleviate the problem of capping the opal during ALD.

Two different types of sacrificial films were here tested on quartz substrates which then were used as surfaces for PS self-assembly, namely polymethyl methacrylate (PMMA) and photoresist (S1813) applied by spin-coating. Organized domains (yellow color) could be seen on most of the 1 μm thick photoresist film inter-mixed with regions of bare islands (light pink color).

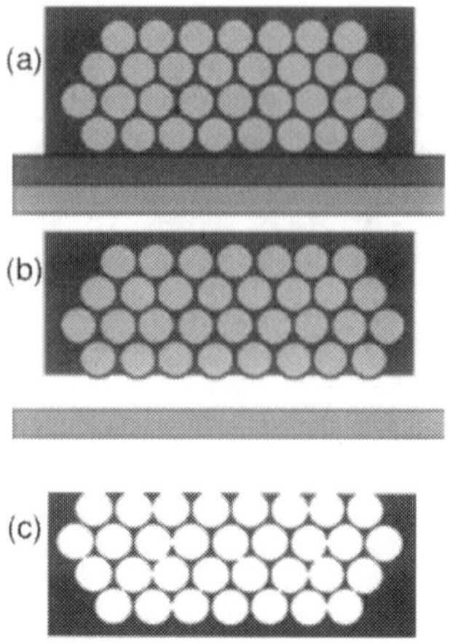

Figure 6. Schematic picture showing the development of a free-standing inverse opal structure grown on a sacrificial template film: (a) infiltration of the fcc opal structure by e.g. atomic layer deposition, (b) removal of the sacrificial layer, and (c) inverse opal structure formed by removal of the embedded PS beads by calcination of solvents, facilitated by the open structure exposed by removal of the sacrificial layer.

In contrast, for the 200 nm thick PMMA film mainly bare substrate surfaces were visible (figure 7), with patches of organized domains (yellow color).

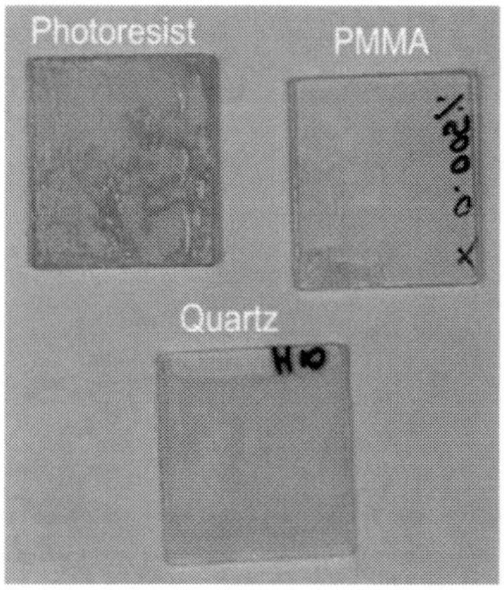

Figure 7. Photographs of opal structures formed on PMMA and photoresist films on quartz, as well as a bare quartz substrate. A PS solution concentration of 0.2 w/v% and a PS bead diameter of 200 nm were employed.

Figure 8 shows optical transmission spectra acquired from domains of self-assembled 200 nm PS beads on 200 nm thick PMMA films (PS concentrations 0.1 and 0.2 w/v%),and 1 μm thick photoresist films (PS concentration 0.2 w/v%), respectively, on quartz. Distinct transmission minima are clearly seen for opals grown on both PMMA and photoresist films, which show that photonic bandgaps form on both types of sacrificial films, albeit over larger surfaces area in the case of photoresist films. This indicates a way forward for easy fabrication of inverse opal structures, where embedded PS beads should be removed after infiltration of the voids in the fcc structure (by e.g. atomic layer deposition or sol-gel techniques), without having to apply etching steps to remove excess infiltration material which other wise inevitably will cap the top of the opal structure (see figure 6).

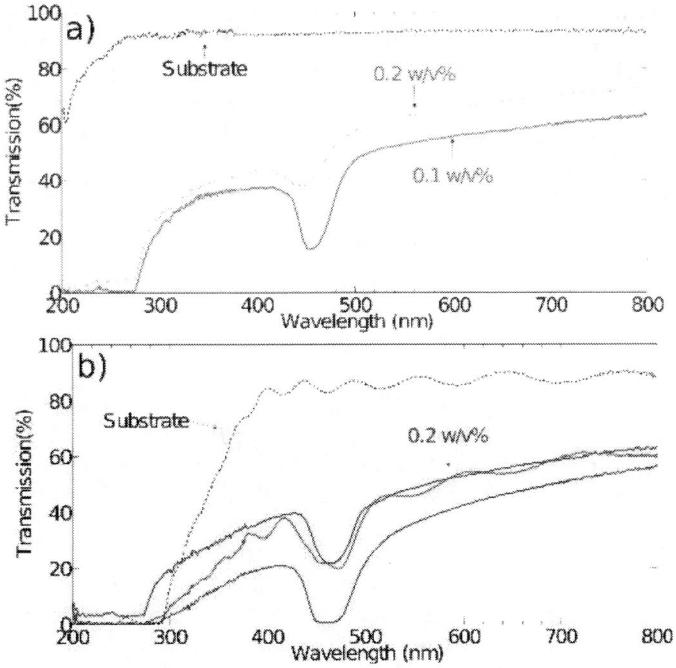

Figure 8. (a) Optical transmission spectra of opal structures grown on PMMA films on quartz (dashed line: substrate; solid line: PS concentration 0.1 v/w%; dotted line: 0.2 w/v%), and (b) Optical transmission spectra of opal structures grown on photoresist films on quartz (dashed line: substrate; solid lines: PS concentration 0.2 w/v% for three different samples.

4. Conclusions

Self-assembly of polystyrene spheres by means of convective evaporation is a viable method to fabricate photonic band gap opal structures. Here we show that in addition to substrate hydrophilicity also surface roughness is important to prepare well-ordered opals. If the substrate is hydrophilic and exhibits a rmssurface roughness < 30% of the bead diameter, an opal structure is obtained. All substrates with these desired characteristics can be used for opal fabrication regardless of substrate morphology and employing a wide range of different types of substrates. A good balance between evaporation rate (regulated via temperature and humidity) and bead concentration should be maintained to obtain substrates completely covered with an opal structure. If the concentration is too low, island growth of the opal leads to formation of opal stripes of beads regularly spaced on the substrate. Finally, we show that it is possible to fabricate well-ordered opal structures on sacrificial PMMA and photoresist films on quartz substrates, with well-developed photonic bandgaps. Our results indicate that more extended opal layers form on photoresist films compared with PMMA films. The latter results open up ways for easy fabrication of open opal structures, which allow for removal of embedded PS after fcc void infiltration, without having to apply etching steps.

5. Acknowledgements

This work was funded by the Swedish Research Council Grant Agreement No. 2010-3514.

The paper was presented at the INERA Workshop "Transition Metal Oxide Thin Films: Functional Layers in 'Smart Windows' and Water Splitting Devices", September 4–6th, 2014, Varna, Bulgaria. The Workshop is part of the Program of the INERA REGPOT Project of the Institute of Solid State Physics, Bulgarian Academy of Sciences.

Journal of Physics: Conference Series **559** (2014) 012007 doi:10.1088/1742-6596/559/1/012007

References

[1] Tarhan I I and Watson G H 1996 *Phys. Rev. Lett.*76 315-18
[2] Wijnhoven J E G J 1998 *Science* **281** 802 - 804
[3] King J S, Graugnard E and Summers C J 2005 *Adv. Mater.* **17** 1010-1013
[4] Li Y, Kunitake T and Fujikawa S 2006 *J. Phys. Chem. B* **110** 13000-4
[5] Cao Y, Wang Y, Zhu Y, Chen H, Li Z, Ding J and Chi Y 2006 *Superlattices Microstruct.* **40** 155-160
[6] Perez N, Huls A, Puente D, Gonzalezvinas W, Castano E and Olaizola S M 2007 *Sensor Actuat B Chem* **126** 86-90
[7] John S 1987 *Phys. Rev. Lett.* **58** 2486
[8] Yablonovitch E 1993 *J. Opt. Soc. Am. B* **10** 283
[9] Yablonovitch E 1994 *J. Mod. Opt.***41** 173-94
[10] Polman A and Wiltzius P 2001 *MRS Bull.* **26** 608-13
[11] Lopez C 2003 *Adv. Mater.* **15** 1679-1704
[12] Chen J I L, von Freymann G, Choi S Y, Kitaev V and Ozin G A 2006 *Adv. Mater.* **18** 1915-19
[13] Goldenberg L M, Wagner J, Stumpe J, Paulke B R and Gornitz E 2002 *Langmuir* **18** 3319-23
[14] Kuai S L, Hu X F, Hach_e A and Truong V V 2004 *J. Cryst. Growth* **267** 317-324
[15] Yannopapas V, Stefanou N and Modinos A 2001 *Phys. Rev. Lett.***86** 4811-14
[16] Rengarajan R, Mittleman D M, Rich C and Colvin V L 2005 *Phys. Rev. E* **71** 016615
[17] Li H l and Marlow F 2006 *Chem. Mater.* **18** 1803-10
[18] Wang X, Husson S M, Qian X and Wickramasinghe S R 2009 *Mater. Lett.* **63** 1981-83
[19] Perez N, Sanza F J, Rodriguez A and Olaizola S M 2011 *Cryst. Res. Technol.* **46** 1044-50
[20] Kim M R, Im S H, Kim Y S and Cho K Y 2013 *Adv. Funct. Mater.* **23** 5700-05
[21] Gu Z Z, Fujishima A and Sato O 2002 *Chem. Mater.* **14** 760-65
[22] Fudouzi H 2004 *J. Colloid Interface Sci.* **275** 277-83
[23] Zhang L and Xiong Y 2007 *J. Colloid Interface Sci.* **306** 428-32
[24] Routh A F and Russel W B 1998 *AIChE J.* **44** 2088-98
[25] Nozar P, Dionigi C, Migliori A, Calestani G and Cademartiri L 2003 *Synth. Met.* **139** 667-70
[26] Lu Y, Yin Y, Li Z Y and Xia Y 2002 *Langmuir* **18** 7722-27

Journal of Physics: Conference Series **559** (2014) 012008 doi:10.1088/1742-6596/559/1/012008

Incorporation of nitrogen into TiO_2 thin films during PVD processes

I Asenova[1, 2, 3], D Manova[2] and S Mändl[2]

[1]Faculty of Physics, Sofia University "St. Kliment Ohridski", 5 James Bourchier Blvd, 1164 Sofia, Bulgaria
[2]Leibniz-Institut für Oberflächenmodifizierung, 15 Permoserstraße, D-04318 Leipzig, Germany

E-mail: iasenova@phys.uni-sofia.bg

Abstract. In this paper we investigate the possibility of incorporating nitrogen into amorphous, photocatalytic TiO_2 thin films, prepared at room temperature, during the growth process. The aim is to reduce the bandgap of the UV active thin films. Physical vapor deposition experiments employing a titanium vacuum arc with gas backfill ranging from pure oxygen to pure nitrogen, are carried out. The resulting films are characterized for chemical composition, phase composition, optical properties and hydrophilicity in order to determine a correlation between gas composition and thin film properties. The experimental results point that a visible change in the band structure of the deposited layers is achieved.

1. Introduction

Titanium dioxide is well known as a photoactive material [1, 2], either as a photocatalyst activated by UV irradiation or by reducing the surface energy under illumination thus exhibiting superhydrophilic [3], self-cleaning or anti-fogging surfaces. Except for powders, TiO_2 thin films can be produced using different ion assisted physical vapor deposition processes where either high temperature or increased ion energy is necessary to obtain an active phase [4]. Temperature sensitive substrates require low temperatures, ideally room temperature where an amorphous photoactive phase can be formed [5, 6]. Independent of the phase composition the band gap is always more than 3 eV (the exact value depending on the polymorph) necessitating UV-A radiation for activation. For increasing the reactivity of the thin films, especially indoors, a reduced band gap is desired. Doping with transition metals or with nitrogen has been reported in the literature [7, 8]. The latter dopant is, theoretically, readily accessible during the physical vapor deposition (PVD) processes. However, the incorporation of nitrogen into the growing film, in contrast to ion implantation into TiO_2 thin films or selective oxidation of TiN, is a much more complex process which is presently not completely understood.

In this report we utilize a one-step physical vapor deposition (PVD) process for obtaining N-doped TiO_2 layers at room temperature, which maintains the usage of temperature sensitive substrates. The deposition is performed for a range of backfill gas composition (oxygen and nitrogen) in order to investigate the influence of the total pressure and nitrogen/oxygen ratio on the film properties. The resulting layers are subsequently investigated for correlation between their chemical composition, optical properties and hydrophilicity versus the gas composition. Secondary ion mass spectrometry

[3] Corresponding author.

(SIMS), scanning electron microscopy (SEM), Raman spectroscopy, UV absorption spectroscopy and contact angle measurement were carried out for films characterization.

2. Experiment

Silicon (100) and fused silica substrates were used for the investigations with fused silica necessary for the transmission measurements. No influence of the substrate on the phase composition was observed in previous work for pure TiO_2 [9]. The samples were positioned at distance of 39 cm from a d.c. cathodic arc with a pure titanium (99.99%) cathode, operating at 100 A. In order to increase the average kinetic energy of the Ti ions emitted from the arc, negative high voltage pulses of 3 kV with length of 30 μs at repetition rate of 3 kHz were applied to the substrate. All the experiments started at room temperature and no additional heating or cooling was employed. The deposition time for all samples was fixed to be 300 s with the substrate temperature at 50 °C or below after the deposition process. Oxygen and nitrogen in different composition ratios were used as a backfill gas with a flow rate ranging between 0 and 60 sccm. The obtained growth rate was around 0.85 nm/s, resulting in film thicknesses of approximately 250 – 280 nm. For comparison pure TiN and TiO_2 thin films were produced using the same deposition parameters. The pressure during the deposition process was always between 0.1 and 0.15 Pa.

Table 1. Oxygen and nitrogen flow rates for PVD thin films deposition.

Sample No.	O_2 gas flow (sccm)	N_2 gas flow (sccm)
TiN	0	50
TiO_2	60	0
TiO30N30	30	30
TiO40N20	40	20
TiO50N10	50	10
TiO40N10	40	10
TiO45N5	45	5
TiO45N15	45	15
TiO40N5	40	5

The film thickness and deposition rate were derived from secondary ion mass spectrometry in a time-of-flight setup (TOF-SIMS) using 15 keV $^{69}Ga^+$ ions for analysis and 2 keV $^{133}Cs^+$ ions for sputtering, both incident at an angle of 45°. The sputter profiling was stopped right at the film-substrate interface to avoid the introduction of errors during the conversion of the time into a depth scale. Measuring the crater depth with an optical interferometer and assuming an identical sputter rate for the different layers leads to the determination of the film thickness. With the aid of the TiO_2 and TiN films produced in the same experiment, the nitrogen content in the layers was derived from the Cs_2N^+ secondary ion signal which showed the lowest matrix effect while still yielding enough signal intensity.

The film morphology was examined with scanning electron microscopy (SEM) using two different acceleration voltages of the electron beam (2 kV and 10 kV). The band gap was determined via ultraviolet (UV) absorption spectroscopy therefore identical layers were deposited on fused silica substrates. In order to investigate the hydrophilicity of the layers, water contact angle measurements were conducted before and after exposing the samples to irradiation for 3 hours. The irradiating UV-A light was generated from an actinic tube with a spectral range of 300 – 460 nm and a maximum at 365 nm, with an incident power density at the substrate position of 1 mW/cm^2. The ambient temperature and humidity were 21 °C and 25% respectively. The contact angle measurements on the irradiated samples were conducted within three hours after the end of the illumination.

3. Results and discussion

As shown by previous studies on the subject [4, 9], titania films obtained via this procedure should be either amorphous or containing nanocrystallites smaller than 5 nm due to the rather low temperature and ion bombardment which is not sufficient for a complete crystallization. Accordingly, XRD and Raman measurements of the films containing an oxygen/nitrogen mixture yielded no information, indirectly confirming the amorphous or nanocrystaline nature.

In all the cases the film thickness was found to be similar (within the measurement error) which indicates a growth rate independent of either the ratio of the backfill gases or the total pressure [10]. Substoichiometric TiO_x is known to be very reactive and adsorbing a large amount of oxygen from the gas phase. Only when the oxygen supply is nearly depleted, i.e. the pressure is decreasing, a reduced oxygen content can be observed. In contrast, variations in the nitrogen gas flow lead to an immediate change in the Ti/N ratio for pure TiN thin films [11].

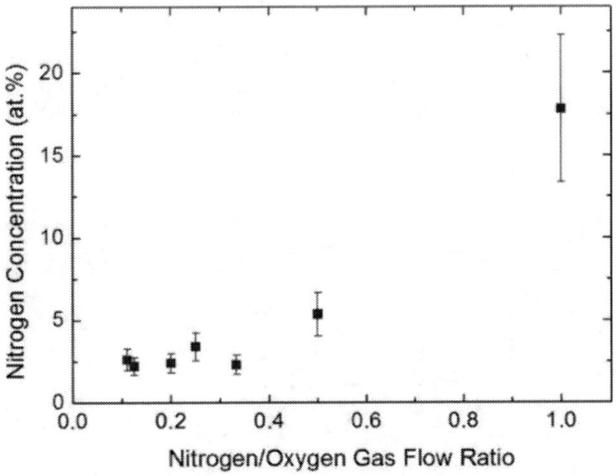

Figure 1. Amount of nitrogen incorporated in the layers versus the flow ratio of the backfill gases. The error bar from the assumed model is included.

The nitrogen concentration in the layers was extracted from the SIMS spectra using the procedure detailed in the experimental section. The results are plotted in figure 1. For all the samples grown with ratio of the backfill gases $N_2/O_2 < 0.3$ the amount of the incorporated nitrogen is the same within the error of calculation – about 3 at.%, if the linear concentration interpolation is valid. When the N_2/O_2 ratio in the gas phase exceeds 0.3, the nitrogen content in the films increases significantly – the sample with the highest amount of incorporated nitrogen of about 18 at.%, produced with an N_2/O_2 ratio of 1:1, implies a stoichiometry of this film corresponding to Ti_3O_2N or $(TiO_2)(TiN)$.

An alternative explanation with the nitrogen content inside the films being near zero for a nitrogen/oxygen gas flow of up to 0.25 cannot be ruled out from the experimental SIMS data as there is no alternative method for determining the nitrogen content for these rather thin films. However, using either interpretation, the competition between reactive oxygen and nearly inert nitrogen is skewing or delaying the nitrogen uptake towards a nitrogen poor regime (a N:O ratio of 1:2 in the gas leads to a N:O ratio of about 1:10 in the growing film). Alternatively, nitrogen radicals or plasma excitation are necessary to increase the nitrogen uptake [12].

The growth structure of the films has been investigated using cross-section examinations by SEM. The results for selected samples are presented in figure 2. All the layers are dense due to the energetic bombardment and the resulting surface mobility during the deposition process, and do not exhibit a porous, defect-rich structure. The viewgraphs of samples TiN and TiO30N30 reveal a distinct columnar morphology [13] with a column diameter of 20 – 30 nm, while the TiO_2 film exhibits less prominent columns of approximately the same width. However, the observed structures are predominantly indicative of the fracture behavior and may not be representative of the film microstructure itself.

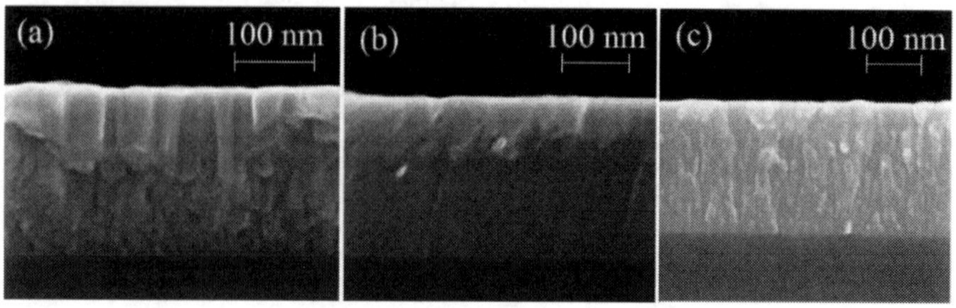

Figure 2. SEM cross-sectional viewgraphs of PVD films: (a) sample TiN; (b) sample TiO_2; (c) sample TiO30N30.

The surface viewgraphs of the same samples show minor droplet contamination from the arc deposition process which consists of metallic titanium, as depicted in figure 3. The lateral dimensions of the macrostructures – as shown in figure 4 – embedded in the layers are about 100 – 300 nm with the elliptical shape indicating a strong compressive biaxial stress within the films, again in agreement with the deposition process as described by general PVD models [14].

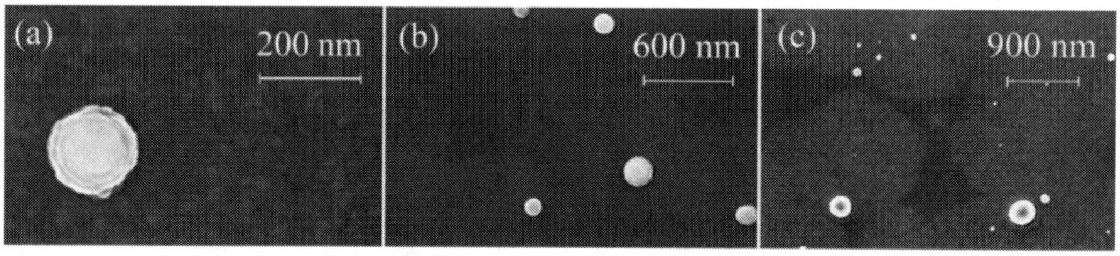

Figure 3. SEM surface topology viewgraphs of PVD films with macroparticles embedded in the film: (a) sample TiN; (b) sample TiO_2; (c) sample TiN30O30.

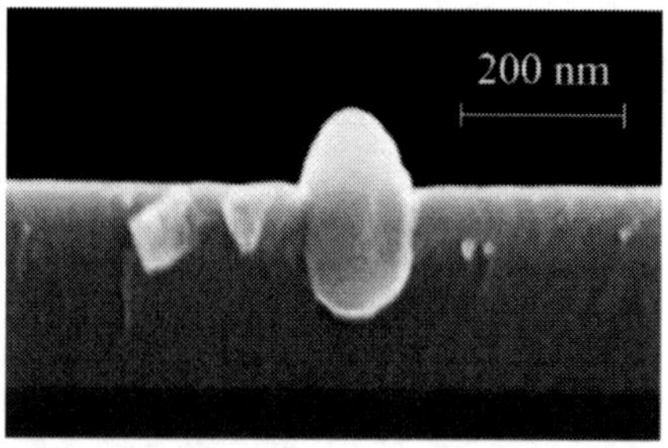

Figure 4. SEM image showing a cross-section of the sample TiO_2 with an embedded macroparticle.

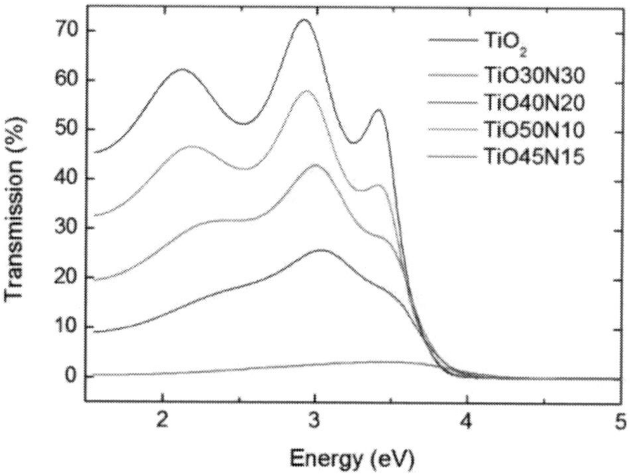

Figure 5. Transmission spectra for selected samples.

Optical properties of selected samples were investigated with UV absorption spectroscopy. The optical band gap E_g of the films was calculated using the Tauc plot formalism for amorphous samples. Here, linear regions for all the layers are observable, with the exception of samples TiN and TiO30N30. The corresponding transmission spectra used for preparing the Tauc plot are shown in figure 5, while the results, combining the values of E_g and nitrogen content, are systematized in table 2. The linear extrapolation of the relevant regions leads to values of $E_g \approx 3.2$ eV for most of the samples, slightly smaller than the optical gap for amorphous TiO_2 (this value for the pure titania film is in agreement with tabulated literature data). For the sample TiO40N20, a band gap narrowing down to $E_g \approx 3.0$ eV is observed. The samples produced with a lower nitrogen gas flow show a significant, smaller shift of the bandgap. In contrast, the coating TiO30N30 is already semimetallic, non-transparent in the optical region.

Table 2. Band gap E_g and nitrogen content of selected samples.

Sample No.	E_g (eV)	E_g (nm)	Nominal Nitrogen Content (at.%)
TiO_2	3.23	384	0
TiO50N10	3.21	386	2.4
TiO45N15	3.15	394	2.3
TiO40N20	3.07	404	5.3
TiO30N30	semimetallic	semimetallic	18

Assuming that the redshift in the band gap is due to the incorporation of nitrogen, a finite, non-zero nitrogen content is present for all samples produced with a nitrogen/oxygen gas mixture. Changes in the phase composition or morphology as the origin of the shift are unlikely. Apparently, a transition is occurring with a lower relative nitrogen uptake for a small nitrogen gas flow and a higher relative nitrogen uptake for higher nitrogen gas flow. The low nitrogen content region corresponds to films with a reduced band gap, while the high nitrogen content leads to semimetallic films.

No information about atomic positions of the nitrogen or influence of the gas flow is available at the present. For crystalline lattices, both interstitial and substitutional N doping of TiO_2 has been reported with the additional formation of detrimental oxygen vacancies [15]. However, the designation of interstitial or substitutional sites in amorphous thin films is not possible.

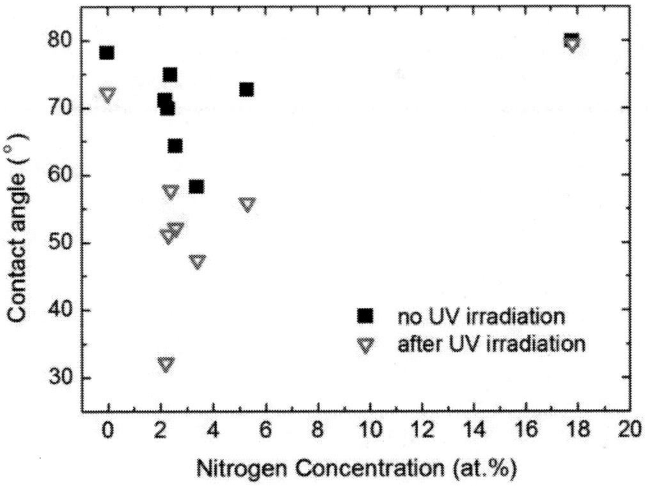

Figure 6. Contact angle of water before and after illumination as a function of the amount of incorporated nitrogen.

In order to test the photoactivity, i.e. the hydrophilicity of the layers, contact angle measurements were conducted before and after the samples being irradiated with UV-A light: the results are displayed in figure 6. With the exception of samples TiN and TiO30N30, which show no difference before and after the illumination, all the layers exhibit a tendency of contact angle reducing as the oxygen backfill decreases. This tendency is more prominent after irradiation, with a minimum found near 3 at.% nitrogen incorporation into the film. For higher contents, the contact angle increases again.

When interpreting these results, it has to be kept in mind that although the present deposition conditions lead to a significant reduction of the contact angle under illumination, the saturation level around 5° is not reached – the relative changes are for the present investigations more interesting than the absolute value. Furthermore, no influence of illumination at reduced energies or increased wavelengths has been investigated up to now.

4. Summary

Amorphous TiO_2:N layers were deposited at room temperature using a PVD method with pulse bias voltage and various backfill gas compositions. A correlation between the backfill gas composition and the nitrogen content and, furthermore, the band gap has been observed. Initially, for small nitrogen gas flows, a low nitrogen reactivity leads to a rather low relative nitrogen incorporation – which nevertheless results in a significant increase in the photoactivity, respective hydrophilicity. For higher gas flows, more nitrogen is incorporated (up to $(TiO_2)(TiN)$ in the present experiment) while the photoactivity under UV-A illumination is reduced again. For these irradiation conditions, only a very small amount of nitrogen incorporation and bandgap shift leads to optimized conditions. At the same time, amorphous or nanocrystalline structures have been observed which should not allow a distinction between interstitial or vacancy incorporation. Additional XPS measurement could be helpful for investigating the electronic structure.

Acknowledgements

This study was partially supported by grant D04-142/28.03.2014 index MU-14-36/2014 from the Ministry of Education and Science of Bulgaria. We would like to thank D. Hirsch and U. Decker for the SEM and optical transmission measurements respectively.

References

[1] Hofmeister W, Tillmanns E and Bauer W H 1984 *Acta Crystallogr. C* **40** 1510
[2] Fujishima A and Honda K 1972 *Nature* **238** 37
[3] Wang R, Hashimoto K, Fujishima A, Chikuni M, Kojima E, Kitamura A, Shimohigoshi M and Watanabe T 1997 *Nature* **388** 431-432
[4] Löbl P, Huppertz M and Mergel D 1994 *Thin Solid Films* **251** 72-79
[5] Schneider J M, Rohde S, Sproul W D and Mattheus A 2000 *J. Appl. Phys.* **33** R173-R186
[6] Manova D, Gjevori A, Haberkorn F, Lutz J, Dimitrov S, Gerlach J W, Valcheva E and Mändl S 2009 *Phys. Status Solidi A* **206** No.1 71-77
[7] Morikawa T, Asahi R, Ohwaki T, Aoki K and Taga Y 2001 *Jpn. J. Appl. Phys.* **40** L561-L563
[8] Lee D, Cho Y, Yi W, Kim T, Lee J and Jung H 1995 *Appl. Phys. Lett.* **66** 815
[9] Gjevori A, Nonnenmacher K, Ziberi B, Hirsh D, Gerlach J W, Manova D, Mändl S 2009 *Nucl. Instrum. Meth. Phys. Res. B* **267** 1658-1661
[10] Thorwarth G, Mändl S, Rauschenbach B 2000 *Surf. Coat. Technol.* **125** 94
[11] Huber P, Manova D, Mändl S, Rauschenbach B 2001 *Surf. Coat. Technol.* **142/144** 418
[12] Matsui H, Tabata H, Hasuike N, Harima H and Mizobuchi B 2005 *J. Appl. Phys.* **97** 123511
[13] Thornton J A 1974 *J. Vac. Technol.* **11** 666-670
[14] Anders A 2010 *Thin Solid Films* **518** 4087-4090
[15] Lin H, Rumaiz A K, Schulz M, Huang C P and Ismat Shah S 2010 *J. Appl. Phys.* **107** 124305

Photocatalytic oxide films in the built environment

Lars Österlund* and Zareh Topalian
Department of Engineering Sciences, The Ångström Laboratory, Uppsala University, SE-751 21 Uppsala, Sweden

E-mail: lars.osterlund@angstrom.uu.se

Abstract. The possibility to increase human comfort in buildings is a powerful driving force for the introduction of new technology. Among other things our sense of comfort depends on air quality, temperature, lighting level, and the possibility of having visual contact between indoors and outdoors. Indeed there is an intimate connection between energy, comfort, and health issues in the built environment, leading to a need for intelligent building materials and green architecture. Photocatalytic materials can be applied as coatings, filters, and be embedded in building materials to provide self-cleaning, antibacterial, air cleaning, deodorizing, and water cleaning functions utilizing either solar light or artificial illumination sources – either already present in buildings, or by purposefully designed luminaries. Huge improvements in indoor comfort can thus be made, and also alleviate negative health effects associated with buildings, such as the sick-house syndrome. At the same time huge cost savings can be made by reducing maintenance costs. Photocatalytic oxides can be chemically modified by changing their acid-base surface properties, which can be used to overcome deactivation problems commonly encountered for TiO_2 in air cleaning applications. In addition, the wetting properties of oxides can be tailored by surface chemical modifications and thus be made e.g. oleophobic and water repellent. Here we show results of surface acid modified TiO_2 coatings on various substrates by means of photo-fixation of surface sulfate species by a method invented in our group. In particular, we show that such surface treatments of photocatalytic concrete made by mixing TiO_2 nanoparticles in reactive concrete powders result in concrete surfaces with beneficial self-cleaning properties. We propose that such approaches are feasible for a number of applications in the built environment, including glass, tiles, sheet metals, plastics, etc.

1. Introduction

World-wide, 30-40% of all primary energy is used in buildings and up to 40% of greenhouse gas emissions come from the building sector, mostly from energy use during the lifetime of buildings [1]. Here, new materials have an important role for future green buildings and for upgrading today's buildings. These materials can be used in a number of devices and in new coating technologies that can be integrated or added into the built environment. Proof-of-concepts have already been demonstrated or even commercialized in a few cases. Examples include controlling the input of visible light and solar energy in smart windows, pollutant abatement of urban (outdoor) air via photocatalysis occurring on exterior building surfaces, self-cleaning surfaces that avoid build-up of contaminants and hence reduces maintenance costs, daylight through light piping systems, hot water or warm air via selective absorption in solar collectors, electricity through building-integrated solar cells, providing coolness via infrared-selective emission towards the clear (night) sky, demand-controlled ventilation by the use of air quality sensors, phase change materials as heat sink and source in construction elements, and superior thermal insulation via nanomaterials (aerogels) based vacuum panels. Materials for these applications have many features in common, not only regarding applications area but also materials types, and are now appearing as an international research field with strong internal cohesion [2].

Energy, comfort and health in the built environment are intimately connected. It is well documented that indoor air quality deterioration is due to occurrence of a large number of chemicals, or classes of chemicals. Among the most important categories of chemicals that occur in the indoor air are volatile organic compounds (VOC). The most commonly found are BTXS (benzene, toluene, xylenes and styrene), terpenes (α-pinene, limonene, etc.) and carbonyls (formaldehyde, acetaldehyde, and acetone). They originate from an increasing number of household products, electronic devices, and synthetic building materials. Simultaneously, changes in building design devised to improve energy efficiency have meant that modern homes and offices are frequently more air-tight than older structures. Whilst these improvements have led to more comfortable buildings with lower running costs, they also provide indoor environments in which contaminants are readily produced and may build up to much higher concentrations than those found outside. Recent reports indicate that adverse health effects are intimately linked to ventilation rates in buildings, although the literature does not provide clear evidence on particular agent(s) for the effects. The limited available data suggest that inflammation, respiratory infections, asthma symptoms and short-term sick leave increase with lower ventilation rates. Home ventilation rates above 0.5 air changes per hour have been shown to be linked to a reduced risk of allergic manifestations among children in a Nordic climate [3]. Higher ventilation rates in offices, up to about 25 l/s per person, are associated with reduced prevalence of sick building syndrome (SBS) symptoms. However, air ventilation is not unproblematic and may be unwanted. It requires energy supply in the heating, cooling, and ventilation (HVAC) system to warm/cool the fresh air, which scales with ventilation rate, and may also cause discomfort.

Surfaces that repel or decompose contaminants such as dirt particles, organic or inorganic molecules or even bacteria, without the need of external labor force or by the use of decontamination agents, are considered to be self-cleaning or easy-clean. It is a concept that has been introduced into different applications in which maintenance free surfaces are desirable [4, 5]. Broadly speaking, and by no means exhaustive, the self-cleaning effect can be divided into three main categories depending on the mechanism by which they operate. These are: (i) Hydrophilic or hydrophobic surfaces, whereby the wetting properties primarily are modified by surface structuring [4], (ii) photocatalytic surfaces where highly oxidizing species attack pollutants and ultimately convert them into CO_2, H_2O and mineral acids aided by light irradiation (preferably sun or indoor air illumination if it is to be a "self-cleaning" action) [5], (iii) surface charging or modification of chemical interactions and surface acid/base properties [6, 7]. Here we present a combination of the latter two strategies for cement surfaces following an approach we recently have demonstrated for TiO_2 thin films on glass substrates, namely surface modification by means of photo-fixation of SO_2 leading to acidic sulfate moieties. These were found to exhibit olephobic or dirt repellent properties as probed by the adherence of a fatty acid [8, 9]. In particular, we show results of beneficial wetting and photocatalytic properties of reactive powder concrete samples with added TiO_2 nanoparticles at various concentrations up to 5 wt% TiO_2.

2. Materials and methods

Concrete samples with various concentrations (0, 1, 3 and 5 wt%) of TiO_2 pigment (Aeroxide P25, Evonik Industries AG) were made by mixing the pigment with reactive powder concrete. The following commercially available materials were used in the concrete sample manufacturing: Nanodur 5941 (Dykerhoff), undensified silica fume (Elkem), Limus 25 (Nordkalk). Glenium ACE (BASF). The amount of Nanodur was adjusted according to replacement level of Aeroxide P25 TiO_2 powders. The use of high performance and ultrahigh performance concrete require the use of large amounts of filler material (size < 125 μm), a low water to cement ratio (w/c) and higher amounts of super-plasticer (SP) than in regular concretes. The w/c ratios can be very low, in the range of 0.18 - 0.25. This requires a different mixing method and duration compared to standard methods, in particular more vigorous and longer periods of mixing. Here, a 1 liter Hobart paddle mixer set at speed "1" was used for the entire mixing sequence. After mixing the dry ingredients, water was added and further mixing was done for 1 min at speed "1". The entire amount of SP was subsequently added and mixing was continued until the dry mass turns into a paste. Rheology measurements of the paste were performed to ensure that it meets the required consistency. The paste was then allowed to "rest", i.e. allow air bubbles to escape from the paste. The paste was cast into standard prism forms (160 mm × 40 mm × 40 mm), and vibrated. The final casted concrete sample were finally covered with plastic and placed in a moist atmosphere at 20 °C.

Samples for further photocatalytic and self-cleaning measurements with dimensions 40 × 40 × 3 mm thick were cut from the concrete cast and polished to obtain a smooth surface. All samples were then left to be equilibrated 50 days in air after manufacturing before further processing. Prior to use all samples were then heated in air at 150°C for 1 h to remove excess water from the manufacturing and then kept in air at room temperature for 1 day before they were used in experiments.

The concrete samples were surface functionalized by means of photo-fixation of SO_2 according to a method which has been described elsewhere [8, 9]. Briefly, the samples were heated to 150°C for 1 h in a gas mixture of air and 50 ppm SO_2 and simultaneously irradiated with UV-A light (λ = 375 nm). This procedure produces SO_2 bonded to oxygen vacancy sites, which after protonation in air yields sulfate terminated TiO_2 surfaces with acidic properties [8, 10]. Figure 1 shows a photograph of the setup used for surface functionalization in this study which allows for functionalization of samples of the order 10 cm × 10 cm.

Figure 1. Setup used for surface functionalization of oxide coated substrates consisting of a chamber with a sample holder and heating plate inside (not visible), gas inlet and outlet, UV transparent window, and UV source consisting of a fan-cooled UV-LED array (λ = 375 nm), which is mounted face-down on the window.

A photocatalyst indicator ink, resazurin (Rz), was used for the rapid assessment of the photocatalytic activity of the concrete samples [11, 12]. It is a commonly used oxidation-reduction indicator in e.g. cell viability assays and as pH indicator, and commercially available as sodium salt. The Rz ink is blue as prepared and irreversibly turns into the pink colored resorufin (Rf) under chemical reduction due to photo-induced reactions (figure 2). In our case the photo-reduction occurs via electron transfer from TiO_2 irradiated with light larger than the band gap, E_g, of anatase TiO_2, i.e. E_g, > 3.2 eV, or λ < 388 nm).

Journal of Physics: Conference Series **559** (2014) 012009 doi:10.1088/1742-6596/559/1/012009

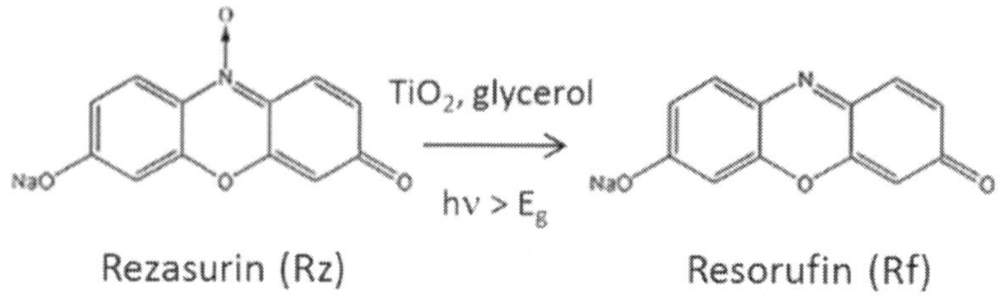

Figure 2. Photo-reduction of resazurin to resorufin by TiO_2.

A thin layer of 0.09 wt% Rz ink in deionized water (Ink Intelligent Ltd., Belfast, United Kingdom) was applied onto the concrete samples using a K-bar, and subsequently dried in air and dark conditions for 24 h. Digital photos of the samples were taken prior to irradiation. The samples were then irradiated with UV-A light (λ = 365 nm, ~10 mW cm^{-2}) for 20 sec, and again photographed. Repeated cycles of UV-A irradiation and photography were performed for a total of 14 min irradiation time, thus yielding a time series of images capturing the photo-induced reactions of the Rz ink on the concrete surfaces. Image analysis was done with the software ImageJ to extract the RGB (red, green and blue) values from the image files. RGB data extraction was carried out on the same sample position with the same number of image pixels for all samples and image series. The blue signal decreases and the red signal increases as the Rz is reduced.

The wetting properties of the concrete samples were acquired by measuring the contact angle of applied droplet of water onto the surface of the samples employing an optical contact-measuring device (OCA15EC, DataPhysics Instruments GmbH, Filderstadt, Germany). The device is equipped with a CCD camera, an automatized syringe control system, and image analysis software. Droplets of water with 5 µl volume were applied onto the surface of the samples by the sessile drop method and digital images were acquired from the top and horizontal view, respectively. A rinsing test was also performed with the samples kept at an angle of 45° to study how water was removed from the surface, as the droplet slides along the tilted surface. In these latter experiments two concrete samples were used: one containing 5 wt% TiO_2 and one which had been further functionalized as described above. In each case 20 µl of water droplets were applied onto the surfaces and images were recorded when the drops reached the end of the samples.

3. Results and discussion

3.1. Photocatalytic activity measurements

Figure 3 shows results from photocatalytic activity test on pure and surface functionalized (S) concrete samples containing 0, 1, 3 and 5 wt% TiO_2. The columns marked "No UV" is photographs before UV-A irradiation, and the columns marked with "UV" depict samples that have been irradiated with UV-A light for 14 min, after which a pink colored areas emerge on samples denoted 1; 1,S; 3; 3,S; 5; and 5,S, respectively. The pink color signals that the samples are photocatalytic active and capable of reducing Rz upon UV-A irradiation. The pure samples (0 and 0,S) show no visible pink color, which indicates that it is the addition of TiO_2 which makes the samples photo-active. Similarly, by inspection of figure 3, it follows that TiO_2 particles are not evenly distributed on the substrate surfaces.

Figure 4 shows the results obtained from the RGB image analysis extraction process performed on the same set of samples as those shown in figure 3. The data are presented as normalized B and R values as a function of UV-A irradiation time. For each data set the R and B values were subtracted by the value at time t = 0, i.e. the initial color of the sample without irradiation to be able to make inter-sample comparisons of the relative color change. It is seen in figure 4 that initially the blue color rapidly increases and that the red color simultaneously decreases during the first few irradiation cycles, depending on sample type.

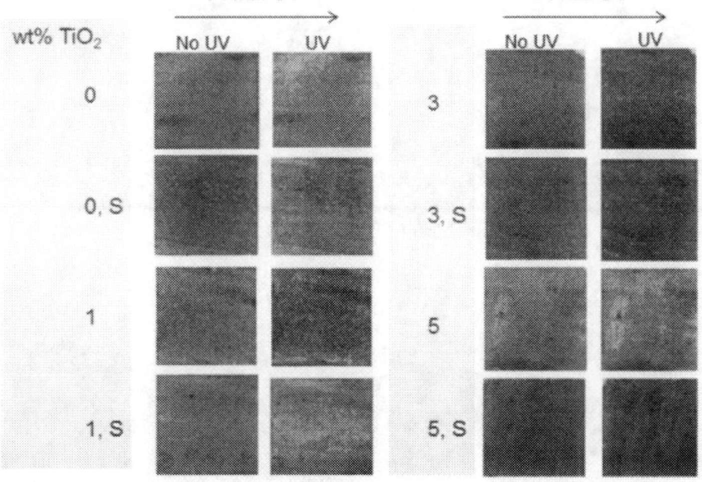

Figure 3. Resazurin dye photocatalytic activity test on pure and surface functionalized (S) concrete samples containing 0, 1, 3 and 5 wt% TiO_2. The samples were irradiated with UV-A light and a total irradiation time of 14 min. Resazurin changes color from blue to pink upon reduction, which is clearly visible on samples containing TiO_2 at concentrations >1 wt% (right panels)

We attribute this to the basic surface properties of the dried concrete samples and the pH sensitivity of Rz, as the concrete surface gradually is re-hydroxylated after the surface functionalization procedure. After extended UV-A irradiation (> 3-5 cycles) the red (blue) color starts to increase (decrease), which shows that the irreversible TiO_2-mediated photo-reduction of Rz to Rf occurs and is the dominant reaction pathway for Rz. The RB-values for both pure and surface functionalized samples with additions of 1, 3 and 5 wt% TiO_2 follow the same trend. The change of the R and B values for the sample 1,S is much smaller than sample 1 without surface functionalization. The difference in RB-values between samples with and without surface functionalization becomes very small at 3 wt% TiO_2, and becomes negligible or even reverse at 5 wt% TiO_2.

This suggests that the photocatalytic activity increases with TiO_2 concentration and that the surface functionalization does not inhibit the activity at high enough TiO_2 concentration ($\geq$ 3 wt%). For concrete samples without TiO_2 (0, and 0,S) a similar RB color change is observed, albeit weak and not visible by direct inspection of figure 3. We tentatively attribute this to the presence of small amounts of photo-active oxide particles in concrete.

Assuming pseudo first-order photocatalytic reaction kinetics of Rz, the average reaction rates for reduction of Rz (blue to pink) obtained over several spots on the concrete sample are determined from RGB color change analysis to be approximately $7 \cdot 10^{-3}$ min^{-1} for both the pure and surface functionalized concrete (without addition of TiO_2 nanoparticles), thus indicating a weak photocatalytic activity. We attribute to small concentrations of photoactive oxide components in concrete. For the concrete samples with additions of TiO_2 nanoparticles the rate is not improved at low concentration of TiO_2 (1 wt%), and after surface functionalization it is even lowered (~ $6 \cdot 10^{-3}$ min^{-1}). At higher TiO_2 concentrations (3 wt% and 5 wt%) the reaction rate increases by 40-60%, which is also manifested in the more distinct color changes (blue to pink) occurring over several parts of the concrete surfaces seen in figure 3.

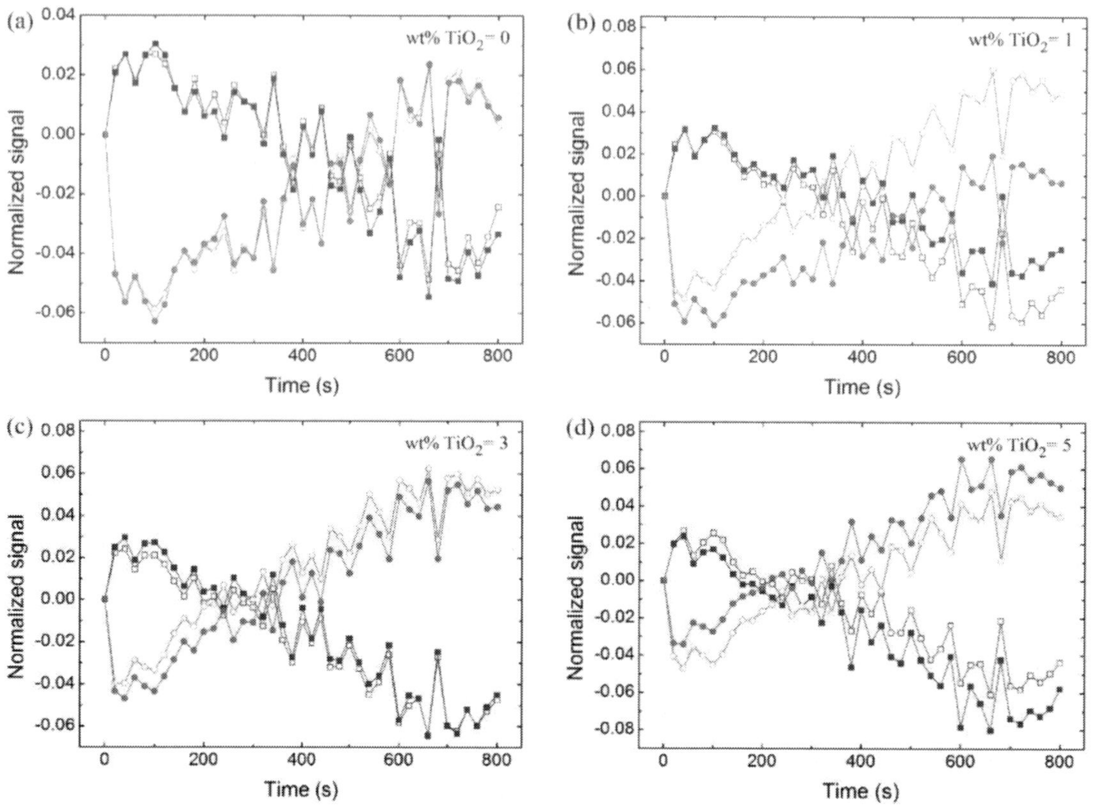

Figure 4. Normalized red (R) and blue (B) color signals as a function of UV-A irradiation time extracted from the images depicted in figure 3. The hollow blue and red symbols (□ and ○) correspond to B and R values of pure concrete samples without addition of TiO_2, and solid symbols (■ and •) the corresponding values for surface functionalized concrete samples containing varying amounts of TiO_2 as indicated in each diagram.

3.2 Wetting measurements

Figure 5 shows results from contact angle measurements on concrete, surface functionalized concrete, concrete with 5 wt% TiO_2, and surface functionalized concrete with 5 wt% TiO_2, respectively. The contact angle for all samples lies between ~10° to 15°, which is characteristic for hydrophilic surfaces, and it is slightly larger for the functionalized samples.

The latter result can be understood from the fact that chemical interactions are mostly dominated by the interfacial acid interactions which influence the strength of the adhesive bonds at the interface and thus control the wetting [13]. The increasing wetting angle on the chemically modified cement surface indicates exchange of protons between neighboring molecules and functional hydroxyl and sulfate groups [7].

Figure 6 shows top view photographs from sessile drop contact angle measurements on concrete samples (a) with 5 wt% TiO_2, and (b) surface functionalized concrete with 5 wt% TiO_2 corresponding to figure 5c and figure 5d, respectively.

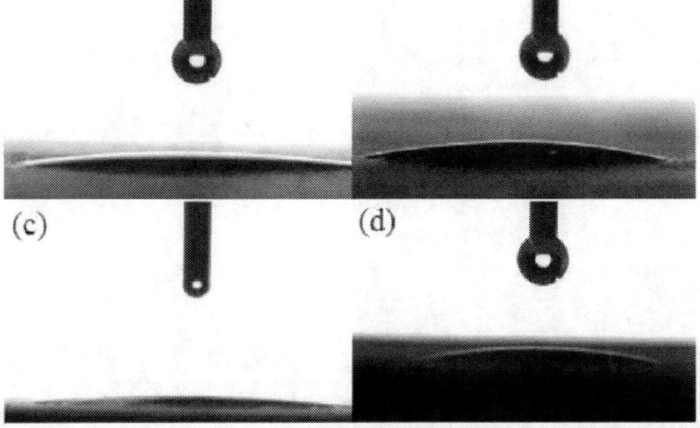

Figure 5. Sessile drop contact angle measurements on (a) concrete, (b) surface functionalized concrete, (c) concrete with 5 wt% TiO_2, and (d) surface functionalized concrete with 5 wt% TiO_2.

In figure 6a it is seen that the water droplet is quickly spread on the sample without surface functionalization after it is applied on the surface, and that it absorbs in the hygroscopic concrete sample over a large surface area. In contrast, on the surface functionalized concrete sample shown in figure 6b, the wetting area is much smaller and stays as a drop on top of the surface for extend times, as water is repelled from the surface and slowly dries (evaporates).

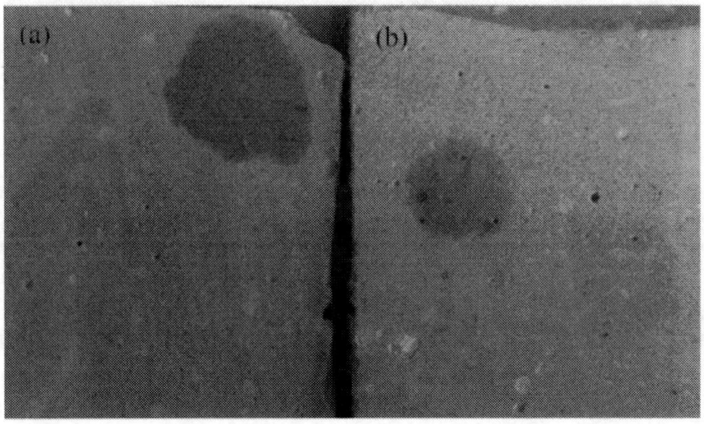

Figure 6. Top view image from sessile drop contact angle measurements on concrete samples (a) with 5 wt% TiO_2, and (b) surface functionalized concrete with 5 wt% TiO_2..

Figure 7 shows results from rinsing test applied to concrete samples containing 5 wt% TiO_2 with and without surface functionalization. Droplets of 20 μl in volume were applied onto the sample surfaces, which were held at an angle of 45°, and the rinsing behavior was noticed as the drop flowed down the surface. It is evident that the droplet flows much more easily on the surface functionalized sample, and that much more water is absorbed on the non-functionalized sample leaving dark moisture

Journal of Physics: Conference Series **559** (2014) 012009 doi:10.1088/1742-6596/559/1/012009

traces. This further shows that the surface functionalization repels water and improves the self-cleaning properties of the concrete surface.

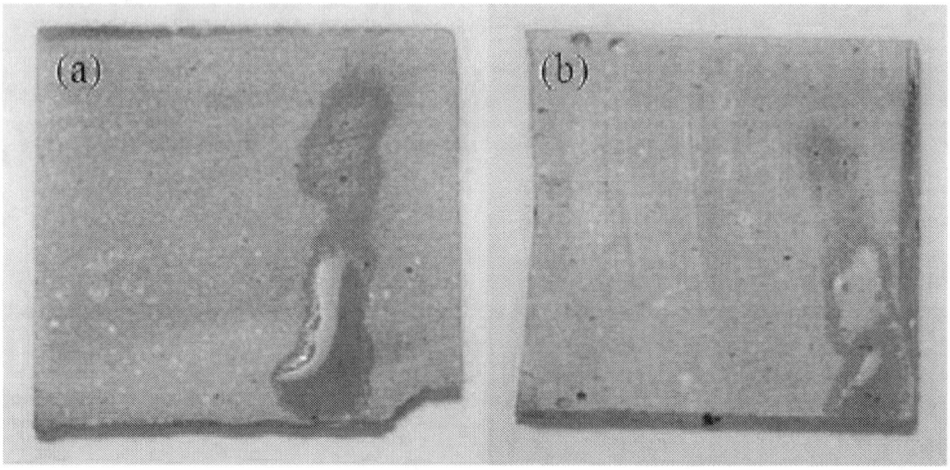

Figure 7. Top view image from sessile drop contact angle measurements on concrete samples (a) with 5 wt% TiO_2, and (b) surface functionalized concrete with 5 wt% TiO_2. The tilting angle of the samples was 45° and the volume of the applied droplet was 20 μl. Water absorbs in the hygroscopic concrete sample without surface functionalization as it slides along its surface. On the surface functionalized concrete sample most water slides off and the trace of absorbed water is much smaller.

The modification of surface acidity can be utilized also to alleviate problems of deactivation in gas-phase photocatalysis. It is well-known that adsorption of reaction products inhibits adsorption of gas pollutants due to strong interaction with surface sites. In particular, carboxylic acids and their ions bind strongly to oxide surfaces such as TiO_2 [6]. This is due to formation of strongly bonded bridging bidentate carboxylate ions. The bonding of formate, which is a common intermediate in degradation of VOCs, is weakened on sulfated TiO_2 and results in considerable improvement of the sustained activity if the surface functionalized photocatalyst compared to untreated anatase TiO_2 [14].

4. Conclusions

We have shown that by mixing TiO_2 nanoparticles with reactive powder concrete and further chemically functionalize the casted concrete sample a surface with enhanced self-cleaning properties is obtained. In particular, the photocatalytic activity and the wetting properties of concrete samples made by mixing TiO_2 nanoparticles (0, 1, 3 and 5 wt% TiO_2) and reactive powder concrete have been measured with and without surface chemical functionalization. Samples containing 1-5 wt% TiO_2, regardless of surface functionalization, show enhanced activity. The surface functionalized concrete samples with 5 wt% TiO_2 concentration showed similar or even higher photocatalytic activity compared to the non-functionalized samples containing the same TiO_2 concentration. All samples – with and without surface functionalization – exhibited hydrophilic wetting properties. The self-cleaning properties as measured by the rinsing properties of applied water droplets at inclined sample angles were found to be superior for surface functionalized samples with added TiO_2. Water was found to slide off the surface of functionalized concrete samples without being absorbed in the hygroscopic, porous concrete surface much more readily than the non-functionalized sample containing the same TiO_2 concentration. We envisage that by adding thin TiO_2 layers on special concrete products, e.g. by spray pyrolysis, or otherwise, a protective and fully functional self-cleaning coating with added surface chemical properties can be obtained following the methods outlined here.

We have previously shown that sulphate surface functionalization of TiO_2 thin film glass leads to dramatically improved oleophobic properties [8, 9], and sustained gas-phase VOC photodegradation activity [14]. It should thus be straightforward to extend the surface functionalization scheme to other surfaces, such as tiles, metal sheets, and plastics. This development would open up possibilities to introduce a second generation of photocatalytic materials in the indoor setting with improved wetting properties combined with photocatalytic activity.

Acknowledgements
This work has been supported by the European Research Council under the European Community's Seventh Framework Program (FP7/2007–2013) Grant Agreement No. NMP4-SL-2013-608950 (SESBE). Dr. Patrick Rogers, CBI, is thanked for preparing reactive powder concrete samples. The paper was presented at the INERA Workshop "Transition Metal Oxide Thin Films: Functional Layers in 'Smart Windows' and Water Splitting Devices", September 4–6th, 2014, Varna, Bulgaria. The Workshop is part of the Program of the INERA REGPOT Project of the Institute of Solid State Physics, Bulgarian Academy of Sciences.

References
[1] United Nations Environment Programme (UNEP) 2007 *Buildings and climate change*
[2] Smith G B, Granqvist C G, *Green Nanotechnology* (CRC Press, Boca Raton, Fl, USA)
[3] Sundell J 2011 *Indoor Air* **21** 191-204
[4] Blossey R 2003 *Nature Mater.* **2** 301-306
[5] Fujishima A, Zhang X, Tryk D A 2008 *Surf. Sci. Reports* **63** 515-582
[6] Davydov A A 2003 *Molecular Spectroscopy of Oxide Catalyst Surfaces* (John Wiley & Sons, Chichester, UK)
[7] John C B (Ed.), 1993 *Wettability* **49** (Marcel Dekker, New York, USA)
[8] Topalian Z, Niklasson G, Österlund L 2012 *ACS Appl. Mater. Interfaces* **4** 672-679
[9] Topalian Z, Niklasson G, Granqvist C G, Österlund L 2009 *Thin Solid Films* **518** 1341-1344
[10] Nisar J, Topalian Z, De Sarkar A, O ̈ sterlund L, Ahuja R 2013 *ACS Appl. Mater. Interfaces* **5** 8516–8522
[11] Mills A, McGrady M 2008 *J. Photochem. Photobiol. A* 193 228–236
[12] Mills A, Wang J. Lee S K, Simonsen M. 2005 *Chem. Commun.* 2721-2723
[13] Fowkes F M, 1987 *J. Adhesion Sci. Technol.* 1 7-27
[14] Topalian Z, Stefanov B, Granqvist C G, Österlund L 2013 *J. Catal.* **307** 265–274

Journal of Physics: Conference Series **559** (2014) 012010 doi:10.1088/1742-6596/559/1/012010

Review on one-dimensional nanostructures prepared by electrospinning and atomic layer deposition

Imre Miklós Szilágyi[1,2*] and Dávidné Nagy[3]

[1]Budapest University of Technology and Economics, Department of Inorganic and Analytical Chemistry, Szent Gellérttér 4., Budapest, H-1111, Hungary;
[2]MTA-BME Technical Analytical Research Group of the Hungarian Academy of Sciences, Szent Gellérttér4., Budapest, H-1111, Hungary;
[3]University of Edinburgh, The King's Buildings, Mayfield Road, Edinburgh, EH9 3JL, UK

E-mail: imre.szilagyi@mail.bme.hu

Abstract. This paper reviews the various 1D nanostructures, which were prepared by electrospinning and atomic layer deposition (ALD). On the one hand, electrospinning served to make sacrificial polymer templates for the ALD growth; and thus various single or multilayer inorganic nanotubes were obtained. On the other hand, polymer, polymer/inorganic or inorganic nanowire templates were produced by electrospinning. By a consecutive ALD reaction various core/shell nanowires were synthesized.

1. Introduction

One dimensional (1D) nanostructures have attracted considerable attention in the recent decades. They have very interesting size-dependent chemical, mechanical, electrical, optical, magnetic, etc. properties, which are due to their one dimensionality. Owing to their unique features, they have been used in a large number of fields, including optoelectronics, nanoelectronics, plasmonics, medical diagnostics, catalysis, drug delivery, therapeutics, separations, and sensing [1-4].

Historically, some decades ago materials synthesis studies focused a great deal to prepare good-quality single crystals. Nevertheless, some materials grow easier in one dimension, rather than in 3D; however, this was considered a problem and the growth of nanowires was tried to be eliminated. The research on nanowires and nanotubes, accelerated in the 1990s, as the nanosized features moved into the center of interest [5-8].

By now, numerous ways have been developed to prepare 1D nanostructures, i.e. spontaneous growth, vapor-liquid-solid (VLS) route, templating on patterned surfaces, growth in porous channels, using self-assembly templates, templating against existing 1D nanostructures, preparation from the vapor phase, solvothermal methods, growth oriented by capping agents, self-assembly of nanoparticles and electrospinning. These are bottom-up approaches, but also top-down methods are available, e.g. isotropic deformation of amorphous or crystalline material, anisotropic etching of single crystals to prepare e.g. grooves, and lithography. These synthesis methods can result in nanorods (with aspect ratios, AR,usually below 20), nanowires (AR>20), nanobelts, nanotubes, nanoneedles, whiskers [5, 9, 10].

Among the various synthesis methods, electrospinning is a convenient and widely used method to obtain 1D nanostructures in a short time and large scale. By this way, polymer, polymer/inorganic and inorganic nanofibers can be prepared. Atomic layer deposition (ALD) is becoming a key tool in nanotechnology, and it can be both a preparation method for 1D nanomaterials and also as a unique technique to program the surface properties of 1D nanosystems.

This paper reviews the various 1D nanostructures, where the synthesis relied both on electrospinning and ALD. Electrospinning served either to make a sacrificial template for the ALD layer to obtain nanotubes, or provided a nanowire template for ALD to prepare core/shell materials.

2. Electrospinning

Electrospinning is a convenient and cost-effective method for the fabrication of nanofibers. The earliest studies date back to the 1930s, but for long the technique remained unexploited. It was newly discovered in the 1990s and has gained a lot of attention since then.

The origin of the method stems from the early "electrostatic spinning" terminology; implying, on the one hand, the application of electrostatic force and, on the other hand, the spinning movement of the ejected polymer solution [11]. The experimental setup basically consists of four main parts: the power supply for achieving high voltage, the needle electrode connected to the polymer solution, the capillary for delivering the precursor solution and the earthed collector (figure 1a) [12].High voltage is applied between the ejection site and the earthed collector. The polymer solution is conducted through a syringe and charged by the applied potential by means of which it starts to form a Taylor cone. The spinning movement is the result of the counteraction of the electrostatic force and the surface tension. By the time the polymer reaches the collector the liquid jet has lost its solvent content and has been elongated due to electrostatic repulsion. The properties of the final product depend upon influencing factors including the characteristics of the polymeric solution, the operation conditions and the surrounding temperature and humidity [13, 14].

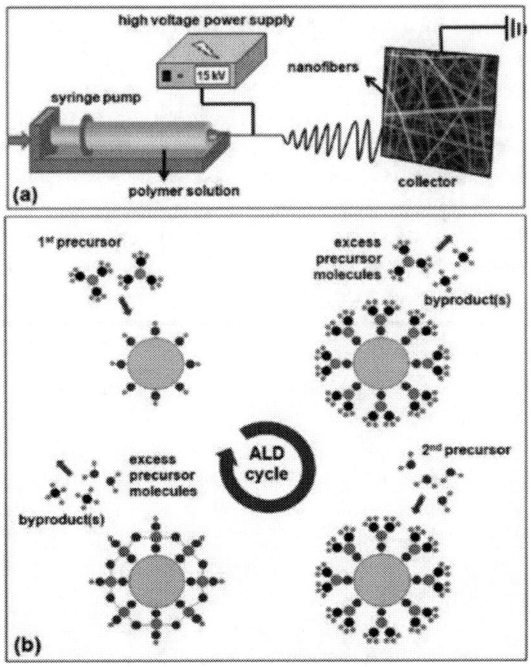

Figure 1. Schematic representation of (a) an electrospinning setup; (b) an atomic layer deposition (ALD) reaction cycle. (Reprinted with permission from [75]. Copyright 2014 American Chemical Society.)

Electrospinning received much attention recently owing to its low cost, high efficiency and good reproducibility [11]. Numerous research areas try to exploit the benefits of what this technique could offer. The nanofibers in general could play key role in environmental protection [15]. In the oil spill cleaning, nanofibrous highly porous membranes are used as sorbents taking advantages of the various functionality, and high uptake capacity could be achieved by electrospun nanofibers [11]. The so-called colloidal nanofibers are considered as good candidates in the field of filtration, adsorption or even as sensors due to the special surface roughness and enhanced properties introduced by the colloidal particles [16]. Besides the high specific surface area, the nanofibers exhibit outstanding charge transport properties owing to their high aspect ratio character, which enables less interfacial interaction of the travelling electrons at grain boundaries. This feature could highly promote the efficiency of light-harvesting devices, such as dye sensitized solar cells (DSSC) or photocatalytic systems [17, 18]. In the opto-electronic industry a specific class of electrospun nanofibers, the conjugated polymer filaments, could potentially revolutionize future devices by providing the attractive possibility to fabricate flexible and light-weight products [13, 19]. Electrospinning attracted substantial interest in the medical research as well, especially in the development of effective drug delivery systems where the high encapsulation and loading efficiency of a wide range of drugs are desired [15].

3. Atomic layer deposition

The basic theory of ALD has been developed already in the 1950s by Aleskovskii in the USSR. He and Koltsov made the first experiments in the 1960s with the method they called molecular layering (ML). Independent of them, Suntola and coworkers developed also the basic principles of ALD in Finland in the 1970s, and it was them who enabled ALD to be used widely in the industry.At first they used only elemental precursors and the method was named atomic layer epitaxy (ALE) [20-23].

The first industrial application of ALD was thin film electroluminescent (TFEL) flat-panel displays. Almost from the beginning, ALD has been researched to be utilized in various other fields as well (e.g. catalysis), but the real international interest has arrived only in the 1990s, when ALD was considered to be a key future method in the semiconductor industry. As the dimensions decreased in silicon-based microelectronics devices, serious conformality problems with the thin films deposited by the existing methods were foreseen to arise in the coming years. Therefore, ALD was selected to be a major method in depositing thin films for several fields, including high dielectric constant gate oxides in metal oxide semiconductor field effect transistors (MOSFETs), copper diffusion barriers in backend interconnects, high aspect ratio trench structures of memory capacitors, etc. Already in the 2000s, major semiconductor companies used ALD partly in their manufacturing technologies. Also in the 2000s, ALD has started to gain interest in nanotechnology, e.g. in sensors, fuel cells, solar cells, catalysis, nanocoatings, displays, LEDs, nanocomposites, nanodevices, etc. [24-28].

A general ALD reaction has four consecutive steps (figure 1b): (i) the first precursor is pulsed into the reactor, which chemisorbs on the surface of the substrate; (ii) the unreacted precursor or reaction byproducts are removed by an inert gas purge and/or by evacuation; (iii) the second precursor is purged, which reacts with the first precursor solely on the surface; (iv) the unused second precursor and the reaction byproducts are removed [29-32].

ALD and chemical vapor deposition (CVD) have many similarities, yet there are several distinctive features, which make ALD unique. In ALD the adsorption of reactants is self-limited and the various precursors meet only on the surface of the substrate [33, 34].

The self-limiting growth leads to the many advantages of ALD. ALD can provide thickness control at an atomic level, as the film thickness can be programmed easily by the number of the ALD cycles. Not only the thickness, but also the composition of the film can be controlled precisely. With ALD, finely tuned doping of thin films or preparation of nanolaminates is easily achievable. ALD can prepare extremely conformal layers, and due to the volatile precursors and self-limiting growth even high aspect ratio structures can be uniformly coated. In ALD there is no shadow effect, which is a hindrance of various other physical and chemical gas phase thin films deposition methods; where, if there is an object between the source and the target, it can reduce the growth rate behind it [35-39].

During the recent decades, ALD precursor chemistry has developed rapidly. ALD can use solid, liquid or gaseous precursors. If the partial pressure of the solid or liquid precursors is low at room temperature, they can be heated to have better sublimation or evaporation conditions. The precursors can be elements, molecules, radicals. By now the variety of materials that can be grown by ALD is very rich. ALD can be used to deposit elements (Au, Ag, Pt, Ru, Fe, etc.), binary compounds (e.g. oxides, nitrides, carbides, sulfides, fluorides, like Al_2O_3, TiO_2, ZnO, HfO_2, GaN, TaC, CaS, SrF_2, etc.), ternary compounds (e.g. $MgAl_2O_4$, TiAlN). Even polymers and organic materials can be grown by ALD. Molecular layer deposition (MLD), a recent version of ALD, where a molecular fragment is deposited during each ALD cycle, can increase the number of ALD prepared materials even further with organic or organic/inorganic (e.g. zincones, alucones) polymers [40-47].

The usual reaction temperature is between 100-400°C, and the ideal reaction temperature zone is often described by the so-called ALD window. The elevated temperature is useful for increasing the precursor and the substrate reactivity, and also to accelerate the sorption processes. However, ALD reactions can be performed at both higher and lower temperatures as well. Low temperature ALD makes it possible to coat heat sensitive substrates, e.g. polymers, biomaterials [48-50].

Considerable efforts have been made to extend the possibilities of traditional ALD to meet the industrial requirements. Plasma or radical enhanced ALD made it possible to deposit Ag or Au, which were not possible by thermal ALD. Usually ALD reactions take place in vacuum, but it is also possible to grow thin films by atmospheric pressure ALD, which makes it much easier to adapt the ALD reaction to other industrial processes. Especially in the semiconductor industry, it might be vital not to grow thin films on the complete substrate but only on parts of it, and this can be achieved by area selective ALD. The slow growth rate of ALD can be overcome by coating a large numbers of substrates at the same time, or by using roll-to-roll or spatial ALD technology [51-54].

As ALD allows deposition of uniform thin films on three dimensional (3D) objects with thickness control of sub-nanometer precision, ALD provides new strategies in modifying the properties of nanoscaled materials and new synthetic routes to novel nanostructures. By now, outstanding results have been achieved by ALD in nanotechnology: preparing or coating nanodots, nanotubes, nanowires, nanolaminates, nanoporous materials, inverse opal structures, various nanocomposites, bionano materials, etc. [55-60].

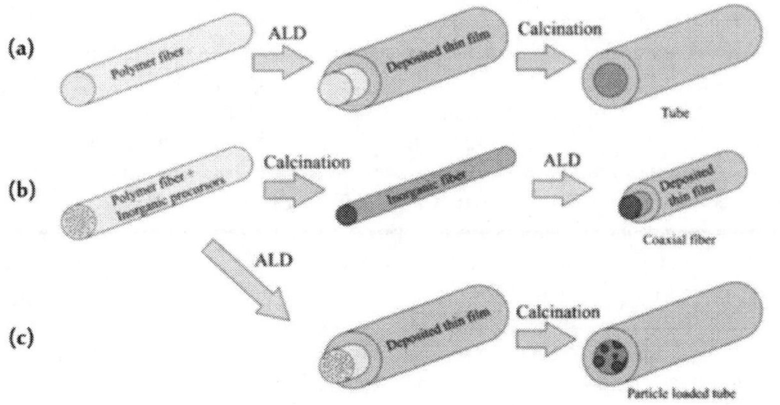

Figure 2. The preparation of (a) nanotubes;(b) core/shell nanowires and (c) particle loaded nanotubes by electrospinning and ALD [61]. (doi: 10.1088/0957-4484/20/3/035602. © IOP Publishing.Reproduced with permission. All rights reserved.)

4. 1D nanostructures obtained by electrospinnning and ALD

The combination of electrospinning and ALD is a powerful tool to prepare nanotubes and core/shell 1D nanostructures, which is clearly reflected by the dozens of already published studies. The various steps for preparing nanotubes and core/shell nanofibers by electrospinning and ALD are summarized in figure 2 [61]. When a core polymer nanofiber is produced by electrospinning, and a shell layer is deposited by ALD, an organic/inorganic core/shell fiber is obtained. If the polymer core is removed by e.g. annealing, a nanotube is prepared (figure 2a). If the electrospinning solution contains inorganic salts as well, then a polymer/inorganic fiber can be electrospun. After annealing the polymer part, the as-obtained inorganic nanofibers can also serve as substrate for the ALD reaction and hence inorganic core/shell nanofibers is produced. It is also possible to load nanoparticles into nanotubes by combining electrospinning and ALD (figure 2b). For this, the as-spun polymer/inorganic nanofiberis covered with an ALD layer before removing the polymer part. If the inorganic salt has appropriate (i.e. low enough) concentration in the electrospinning solution, then instead of a core/shell nanofiber a particle-loaded nanotube will be the product (figure 2c).

In the next sections, at first nanotubes, then core/shell nanofibers are reviewed, which have been obtained by the synergy of electrospinning and ALD.

4.1. Inorganic nanotubes prepared by electrospinning and ALD

Al_2O_3 nanotubes with tuned wall thickness represent one of the first examples of 1D nanostructures, which were produced by combining electrospinning and ALD. 200-400 nm thick polyvinyl alcohol (PVA) nanofibers were electrospun, and Al_2O_3 nanofilms were deposited on the fibers by ALD (60°C, trimethyl aluminium (TMA) and H_2O as precursors). After removing the polymer core by annealing at 400 °C for 24 h, Al_2O_3 nanotubes were obtained with tuned wall thickness (14-28 nm) [62].

Al_2O_3 nanotubes with porous, coral-like structure and extremely high specific surface were produced using the same precursors, but with a different polymer core removal process (figure 3). PVA nanofibers (340 nm thick) were obtained by electrospinning. They were covered uniformly by 25, 50 and 100 nm Al_2O_3 films by ALD. The PVA core was not removed by annealing, but by dissolution in water at varying temperatures and times (20-100°C and 30-180 min, respectively). From 40°C, the polymer swelled in water causing cracks on the shell Al_2O_3 layer. As the dissolved polymer migrated out, a complex channel structure was formed on the tube walls, resembling a nanocoral structure. The as-formed porous nanotubes grew in diameter (1300 nm), and the material had very high specific surface, i.e. 323 m^2/g. This structure was suggested to be beneficial to prepare porous materials for filtration or catalysis [63].

PVA electrospun nanofibers (300 nm thick) served as a template also for a multilayer Al_2O_3 (18 nm)/ZnO (8.5-39 nm)/Al_2O_3 (22 nm) nanotubular structure. Al_2O_3 (TMA and H_2O) and ZnO (diethyl zinc (DEZ) and H_2O, 45 and 65°C, respectively) were deposited by ALD. The PVA core was removed by annealing at 450°C for 12 h, or by dissolution in H_2O for 12 h. After this the Al_2O_3/ZnO/Al_2O_3multilayer nanotubes were annealed at 700°C for 12 h, and during that ZnO diffused into both the inner and the outer Al_2O_3 layers. The vacancy supersaturation and Kirkendall void production resulted in two isolated $ZnAl_2O_4$ spinel nanotubes. The Kirkendall gap between the tube-in-tube structure was influenced by the intermediate ZnO layer [64].

ZnO nanotubes were prepared using polyvinyl pirrolidone (PVP) nanofibers as sacrificial templates as well, but in this case the obtained nanotubes were not multilayer, but single. 180 nm ZnO layers were grown by ALD (70°C, DEZ and H_2O) onto the 630 nm thick PVP nanofibers, then the PVP core was eliminated by annealing at 500°C for 4 h. The ZnO layer consisted of 30 nm ZnO nanoparticles, and the calcinations lead to their oriented grain growth. As a result the wall of the free standing ZnO nanotubes were composed of 200 nm ZnO nanoplatelets [65].

Besides PVP, also polyacrylnitrile (PAN) as-spun nanofibers were templates for ZnO nanotubes. On the 100-150 nm electrosun PAN nanofibers 10, 30 and 50 nm ZnO films were deposited by ALD (150°C, DEZ and H_2O). If the PAN/ZnO composite was only annealed at 500°C for 2h to remove PAN, the thin ZnO walls collapsed. Thus, an O_2 plasma treatment was applied before the annealing, and the nano tubular structure was well preserved. The ZnO nanotubes were tested as gas sensors to ethanol (EtOH). The ZnO nanotube with 10 nm wall thickness was wholly depleted, and due to this it was the most sensitive to 100 ppm EtOH with an extreme response of 1184, and subsecond response time between 350-500°C. The sensor was selective to EtOH, relative to H_2 [66].

Using PAN nanofibers as core materials, gas sensitive SnO_2 nanotubes were obtained as well. 100-200 nm thick PAN nanofibers were electrospun, onto which 8, 22 and 37 nm SnO_2 layers were grown by plasma enhanced (PE) ALD (100°C, dibutyltindi acetate (DBTDA) and O_2 plasma). The amorphous as-deposited SnO_2 became crystalline during annealing (700°C, 1h), and the polymer core was eliminated as well. Similar to ZnO nanotubes, the SnO_2 nanotubes with the smallest wall thickness had the highest response (188) to 100 ppm EtOH and shortest response time (few seconds). This was due to the complete electron depletion and effective gas diffusion through the permeable wall, composed of interconnected SnO_2 nanoparticles. The sensors were selective to EtOH in the presence of reducing gases (H_2, CO, NH_3), but not in the case when oxidizing gases (NO_2) were present [67].

The same group prepared a continuously aligned and assembled structure, i.e. a ca. 50 µm thick yarn, by electrospinning, aligning and twisting PAN nanofibers. They deposited SnO_2 films by the above mentioned PE-ALD process on the PAN yarn then annealed the composites at 700°C. Hence, the obtained SnO_2 yarn consisted of nanotubes with a diameter and wall thickness of 500 and 70 nm, respectively (figure 4). The main advantage of the yarn was its easy handling, which was demonstrated in preparing a H_2 sensitive gas sensor [68].

TiO_2 nanotubes were prepared as well by electrospinning and ALD. 500 nm PVP fibers were electrospun, onto which 60 nm TiO_2 was deposited by ALD (70°C, $Ti(O_iPr)_4$ and H_2O). To remove the polymer, the PVP/TiO_2 fiber was annealed at 500°C for 4h. During this the as-grown amorphous TiO_2 crystallized into 16 nm anatase platelet grains, parallel to the direction of the tube axis [69].

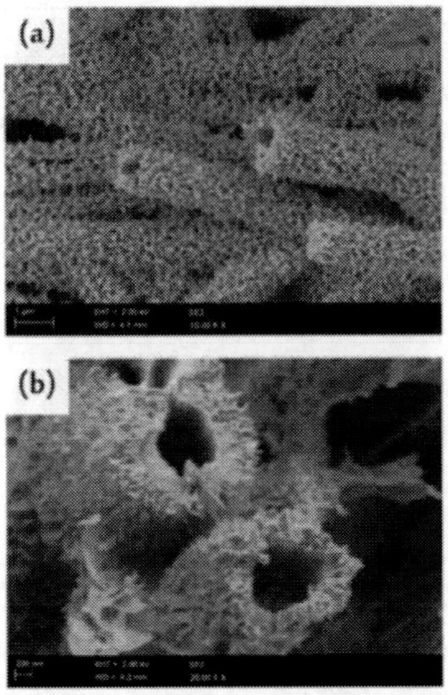

Figure 3.SEM images of Al_2O_3nanocoral formed after dissolving the PVA core from PVA/ Al_2O_3 core/shell nanofibersat (a) lower and (b) higher magnifications. (Reprinted from [63] with kind permission of Springer Science and Business Media.)

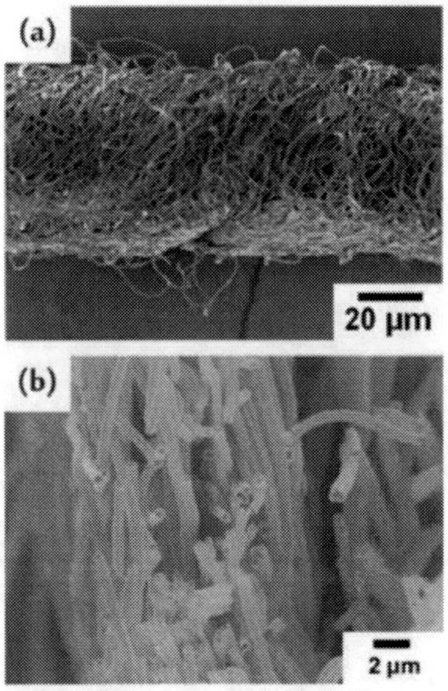

Figure 4. SEM images of (a) SnO_2 nanotube yarn; (b) individual SnO_2 nanotubes [68]. (doi: 10.1088/0964-1726/20/10/105019. © IOP Publishing. Reproduced with permission. All rights reserved.)

TiO_2 nanotubes served a basis of an Ag/Al_2O_3/TiO_2 nanotube composite too. 15-80 nm Ag nanoparticles (NPs) were obtained by sol-gel method and mixed in a polyethylene oxide (PEO) solution. From this Ag NP embedded PEO nanofibers (650 nm thick) were electrospun. Ca. 0.15-1.5 nm Al_2O_3 layer was deposited on the fibers by PE-ALD (65°C, TMA and O_2 plasma), then 4-14 nm

TiO_2 was grown by PE-ALD ($Ti(O_iPr)_4$ and O_2 plasma). The PEO polymer was dissolved in water and the $Ag/Al_2O_3/TiO_2$ nanotube was annealed at 500°C for 1 h. This hybrid system was tested in a DSSC applications), and showed an enhanced light absorption and an increased photocurrent generation, compared to Ag/TiO_2 nanotubes. The plasmonic effect of Ag NPs could be controlled with the thickness of the dielectric Al_2O_3 spacer between the Ag NPs and the dye molecules [70].

SnO_2/TiO_2 multilayer nanotubes were prepared as lithium ion battery anodes. PAN nanofibers (200-300 nm thick) were electrospun, onto which 12 nm SnO_2 and 22 nm TiO_2 layers were deposited consecutively by PE-ALD (100°C, dibutyltindi acetate and O_2 plasma, $Ti(O_iPr)_4$ and O_2 plasma, respectively). The PAN core was removed by annealing at 700°C for 1 h. While pure SnO_2 nanotubes disintegrated into nanoparticles during electrochemical charging/discharging, the TiO_2 encapsulation of SnO_2 prevented this. The high capacity of SnO_2 and the superior cycling performance of TiO_2 were synergistically combined in the SnO_2/TO_2 double shell nanotube electrode [71].

HfO_2 is another material from which nanotubes were produced by electrospinning and ALD. The core polymer nanofibers (70 or 330 nm) were made of nylon 6,6. The fibers were covered by 15 or 70 nm HfO_2 layers by ALD (200°C, $Hf(NMe_2)_4$ and H_2O). The polymer core was removed by heating at 500°C for 2 h to obtain HfO_2 nanotubes [72].

4.2. Organic core/inorganic shell nanofibers produced by electrospinning and ALD

In a pioneer study, the etching of electrospun nylon nanofibers by the most common ALD precursor, i.e. TMA, was investigated, and also it was studied how it could be prevented. Nylon nanofibers were obtained from polyamide-6 (PA-6) solution. Al_2O_3 (60 and 90°C, TMA and H_2O), and ZnO (60°C, DEZ and H_2O) were grown on the fibers by ALD, while an organic/inorganic hybrid layer (120°C, TMA and gylcidol) by MLD. The TMA precursor is highly reactive and it caused significant fiber degradation. When ZnO was deposited prior to Al_2O_3, it decreased the corroding effect of TMA, yet a certain part of TMA could still reach the core polymer by diffusing through the polycrystalline ZnO. In contrast, protecting the fibers by both the ZnO and the MLD layer was effective and the shape of the original polymer fibers was maintained [73, 74].

Electrospun nylon 6,6nanofibers (80, 240 or 650 nm thick) were used also as templates for 90 nm ZnO nanolayers grown by ALD (200°C, DEZ and H_2O). The electrospun nylon/ZnO fibers formed a flexible and mechanically stable mat. Their photocatalytic activity increased with a decrease in the core polymer diameter [75].

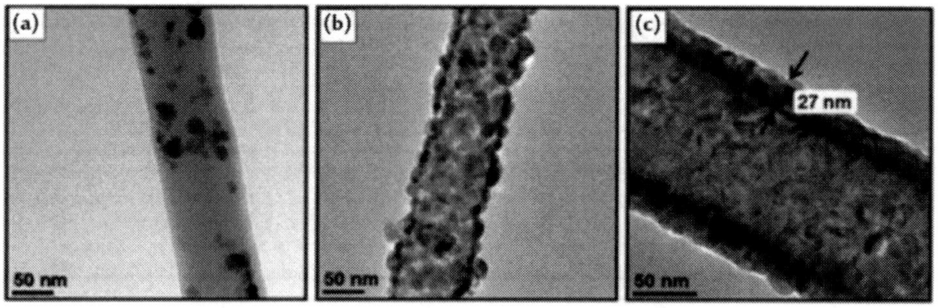

Figure 5. TEM images of electrospun nylon nanofibers coated (a) with remote ZnO nanoparticles (NPs); (b) densely with ZnO NPs; (c) with 27 nm thick layer of ZnO NPs. (Reproduced from [76] with permission of The Royal Society of Chemistry.)

The photocatalytic activity of ALD deposited ZnO (200°C, DEZ and H_2O) nanoparticles or nanofilms on similar, electrospun 80 nm thick nylon 6,6 nanofibers was investigated further. Varying the ALD parameters, i.e. normal dose and dynamic vacuum, increased dose and dynamic vacuum, normal dose and static vacuum, the polymer nanofibers were decorated with remote 20 nm ZnO NPs, with densely packed ZnO NPs, or with a continuous 27 nm thick ZnO layer consisting of ZnO NPs (figure 5). All composites were good photocatalysts, and the nylon nanofibers packed densely with ZnO NPs had the largest activity, due to their highest specific surface [76].

The application of ALD ZnO coated electrospun fibers was studied further in photocatalysis. 1000 nm thick polysulfone (PSU) as-spun fibers were coated by ALD with distinct 10 nm ZnO nanoparticles (quantum dots), and also with 43, 56 and 75 nm ZnO nanocoatings built up by this particles. The PSU/ZnO composite with the thickest ZnO layer had the highest photocatalytic activity. The role of oxygen vacancies (V_Os) and Zn interstitials (Zn_i) were investigated. As the ALD cycle number and the ZnO thickness increased, the number of V_Os grew, while the density of Zn_i remained basically the same. It was concluded that V_Os are more effective in the context of photocatalysis than Zn_is and related defects [77].

In a recent study,a flexible mat of 655 nm thick electrospun PAN nanofibers was covered with 60 nm ZnO thin films by atomic layer deposition. This ZnO layer served as seed in a consecutive hydrothermal process, where 25 nm thick and 600 nm long single crystalline ZnO nanowires were grown on the PAN/ZnO composite fibers. The as-obtained hybrid had excellent photocatalytic activity, which was due to the synergy of the catalytic activity at surface defects (of ALD ZnO seed coating), of the valence band and the conduction band (of ZnO nanoneedles) [78].

Besides ZnO and Al_2O_3, TiO_2 nanofilms were also deposited on electrospun polymer fibers. It was studied how the precursors-substrate interactions during ALD coating influenced the mechanical properties of the fibers. Al_2O_3, ZnO and TiO_2 thin films were deposited by ALD (50 and 150°C) on nylon nanofibers (200 nm in diameter) prepared by electrospinning. Smaller ALD precursors (TMA for Al_2O_3, $TiCl_4$ for TiO_2) diffused into polymer and formed particles there, while larger precursors (DEZ for ZnO and $Ti(O_iPr)_4$ for TiO_2) rather formed a film on the surface of the polymer. The diffusion of oxide precursors into the polymer increased Young's modulus and decreased the ultimate strain [79].

Such nanofibers were prepared by electrospinning as well, which contained both PAN and an iron salt (Fe phtalocyanine). Onto these, TiO_2 was deposited by low temperature ALD (50°C, $TiCl_4$ and H_2O), and the hybrid was pyrolyzed in N_2 at 900 °C for 2 h. As a result core/shell Fe-containing carbon/TiO_2 composite fibers were obtained. These were used as electrocatalysts for the oxygen reduction reaction (ORR) of fuel cells. Adding Ti species to the Fe-C fibers increased the mass activity of the ORR and decreased the production of the side product H_2O_2, i.e. increased the selectivity of the reaction [80].

Al_2O_3 coated Si/C nanofibers are another examples, where not polymer but carbon fibers were in the core. 30-50 nm Si nanoparticles were mixed with PAN, and 150-300 nm Si/PAN composite fibers were prepared by electrospinning. The composite was carbonized by annealing at 800°C for 2 h in Ar atmosphere. Onto the Si/C nanofibers 7-28 ALD cycles of Al_2O_3 were deposited (120°C, TMA and H_2O). The Si-C/Al_2O_3 fibers were tested as anode materials for rechargeable Li ion batteries. The Al_2O_3 coating significantly increased the capacity retention and the coulomb metric efficiency, because it increased the mechanical integrity of the electrode structure and prevented side reactions between the electrode and the electrolyte [81].

4.3. Inorganic core/shell nanofibers produced by electrospinning and ALD

In the last group of composite nanofibers, there is no organic material in the core of the fibers. WO_3/TiO_2 core/shell nanofibers with enhanced visible light photocatalytic activity are good examples for these materials. Polymer/inorganic PVP/ammonium metatunsgstate composite nanofibers were electrospun. They were annealed at 500 °C for 1 h in air to get 250 nm thick WO_3 nanofibers, composed of 20-60 nm interconnected nanoparticles. The authors studied in detail the effect of heating program during annealing and concluded that the polymer/inorganic fibers had to be annealed very slowly (1°C/min) and at the lowest possible temperature, in order to prevent the disintegration of the as-formed oxide fibers into particles [82].On the WO_3 nanofibers 1.5-20 nm TiO_2 thin films were deposited by ALD (250°C, $TiCl_4$ and H_2O) (see figure 6). Under visible light, the WO_3/1.5 nm TiO_2

nanofibers had the highest photocatalytic activity, even higher than Degussa TiO_2, while the thicker TiO_2 layers filled the voids between the NPs of the nanowires and reduced the specific surface area, weakened the photocatalytic activity [83].

The photocatalytic activity of TiO_2 nanofibers was also increased by doping with Nb and depositing Pt nanoparticles onto them. A solution of polyvinyl acetate (PVAc), $Ti(O_iPr)_4$ and $Nb(OEt)_5$ was electrospun, then annealed at 500°C in air for 24 h. As a result, 152-220 nm Nb-doped TiO_2 nanofibers were obtained. Onto them 4-14 nm Pt nanoparticles were deposited by ALD (270°C, $(MeCp)Pt(Me)_3$ and O_2). The hybrid fibers were studied as possible fuel cell anodes. The Nb doping increased the conductivity of TiO_2, and thus the ORR became 20 times higher. A larger Pt load and post-treatment in 5% H_2 at 500°C increased the ORR even further [84].

In another study, it was investigated what happened if the core and shell materials were exchanged in photocatalytic core/shell nanofibers. PVP/$Ti(O_iPr)_4$ and PVP/ZnAc nanofibers were electrospun, then annealed at 500°C in air for 3 h. Then they were covered with ZnO (200°C DEZ and H_2O) and TiO_2 (200°C, $Ti(NMe_2)_4$ and H_2O) thin films by ALD to get 200 nm thick TiO_2/ZnO and 270 thick ZnO/TiO_2 core/shell nanofibers, respectively. These two combinations exposed electrons and holes selectively to the environment, which migrated to TiO_2 and ZnO, respectively. During photocatalysis, the TiO_2/ZnO had considerable higher activity owing to the efficient hole capture by oxygen vacancies, and to the lower mobility of holes [85].

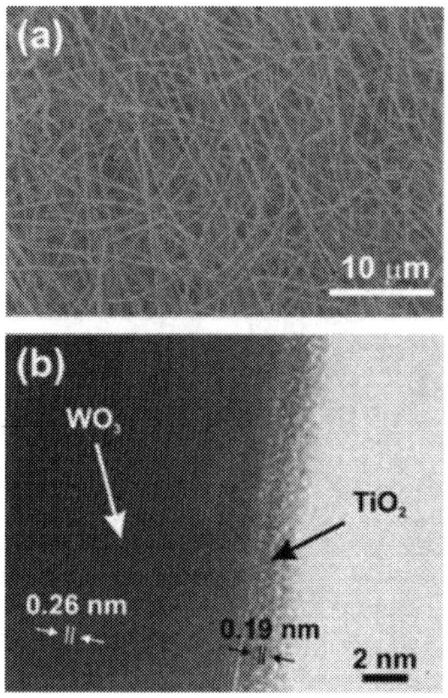

Figure 6. (a) SEM image of electrospun WO_3 nanofibers; (b) HRTEM image of the WO_3/1.5 nm TiO_2 core/shell nanofiberscomposite [83]. (Copyright Wiley-VCH Verlag GmbH & Co. KGaA. Reproduced with permission.)

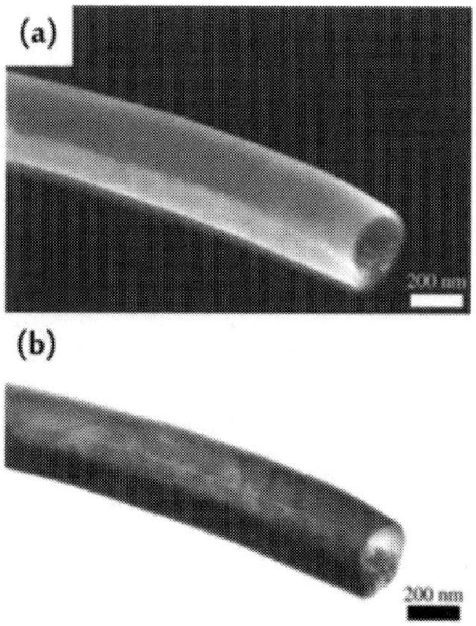

Figure 7. FESEM (a) and STEM (b) images of Fe_2O_3 particle filled TiO nanotube [61]. (DOI: 10.1088/0957-4484/20/3/035602. © IOP Publishing. Reproduced with permission. All rights reserved.)

Reusable magnetic and photocatalytic $CoFe_2O_4$ and Fe_2O_3nanoparticle loaded hollow TiO_2 nanofibers are presented as last examples (figure 7). PVP/$Fe(NO_3)_3$/$Co(NO_3)_2$ and PVP/$Fe(AcO)_2$ polymer/inorganic composite fibers were obtained by electrospinning, onto which TiO_2nanofilms were grown by ALD (250°C, $TiCl_4$ and H_2O). The composites were annealed in air at 500 °C for 4 h to remove the polymer. If the concentrations of the Fe and Co precursors were low enough in the electrospinning solutions, then after annealing the $CoFe_2O_4$ and Fe_2O_3filled only partially the voids of the TiO_2 shells, and nanoparticle loaded nanotubes were prepared. The $CoFe_2O_4$/TiO_2 and

Fe_2O_3/TiO_2composites were good photocatalysts, and after the photocatalytic reaction they could be collected simply by a magnet instead of centrifuging or filtration [61].

5. Conclusion

Electrospinning is an easy and cost-effective way to prepare various nanofibers. ALD has been proven to be a valuable tool not just in semiconductor industry but also in many fields of nanotechnology. Their combination is a powerful way to prepare nanotubes and core/shell nanofibers with tuned properties. The first study on using both electrospinning and ALD to prepare 1D nanostructures was published less than ten years ago, and since then the number of materials prepared by the synergy of these two techniques has increased rapidly.

For electrospinning, such polymers (PVA, PVP, PAN, PEO, nylon, PSU, PVAc) were used, which had some heteroatoms that could serve as nucleation centers for ALD. Onto them usually the most widespread ALD oxide materials (Al_2O_3, ZnO, TiO_2) were deposited, but other oxides (SnO_2, HfO_2) and even noble metals (Pt) were also grown, either by thermal ALD or by PE-ALD.

To obtain nanotubes, the polymer core had to be removed, which was done mostly by annealing, but occasionally also by dissolution or by O_2plasma treatment. In several cases, simple nanotubes (e.g. Al_2O_3, ZnO, TiO_2, SnO_2, HfO_2) were prepared, but also more sophisticated structures (nanotube Al_2O_3yarns, multilayer $Al_2O_3/ZnO/Al_2O_3$, SnO_2/TiO_2 and Ag nanoparticle embedded Al_2O_3/TiO_2 nanotubes) were synthesized.

In the case of many core/shell nanofibers, the as-spun polymer fiber itself served as the core. In many examples, flexible core/shell fiber mats were obtained, which had beneficial mechanical properties. It was shown that depending on their properties, the ALD precursors can diffuse either into the fiber and start nucleation there, or they start to form films only on the surface. By using various precursors and ALD reaction parameters, on the polymer cores Al_2O_3, ZnO and TiO_2 films, and ZnO nanoparticles were grown by ALD, but also hybrid organic/inorganic Al-O-glycidol layers by MLD.In few cases, the polymers contained Fe or Si salts or nanoparticles, and when after the ALD reaction the polymer core was carbonized, and Fe-C/TiO_2 and Si-C/TiO_2 core/shell composites were produced.

When the polymer core was oxidized before the ALD run, inorganic core/shell nanofibers were prepared. When the core inorganic fiber consists of nanoparticles, the ALD layer should be thin in order not to reduce the specific surface by filling the voids between the particles, as was shown by the example of WO_3/TiO_2 composite. When the inorganic salts had lower concentration in the polymer solution, even magnetic nanoparticle loaded hollow Fe_2O_3/TiO_2 or $CoFe_2O_4/TiO_2$ nanofibers could be obtained. With ALD and electrospinning it could be studied what happens if the core and shell materials are exchanged (ZnO/TiO_2, TiO_2/ZnO). Not just films but also nanoparticles could be deposited on the inorganic nanofibers by ALD, as in the case of Nb-doped TiO_2/Pt composites.

The various nanotubes and core/shell nanofibers had versatile applications ranging from photocatalysis and solar cells to batteries, gas sensors and fuel cells. It is clear that the combination of electrospinning and ALD can provide both simple and sophisticated 1D nanostructures as well. The present examples are only to show up the potential in the synergy of these two methods, but there is large room and also need for novel 1D nanomaterials with programmed properties, which predicts a huge development of the field.

6. Acknowledgements

I. M. S. thanks for a János Bolyai Research Fellowship of the Hungarian Academy of Sciences and an OTKA-PD-109129 grant.This paper was presented at INERA Workshop "Transition metal Oxide Films-functionalk layers in " Smart windows" and Water Splitting Devices", September 4-6th, 2014, Varna, Bulgaria. The Workshop is part of the Program of INERA REGPOT Project of Institute of Solid State Physics, Bulgarian Academy of Sciences.

References

[1] Yuan J Y and Muller A H E 2010 *Polymer* **51** 4015

[2] Ye J F and Qi L M 2008 *J Mater Sci Tech* **24** 529
[3] Wernersson L E, Thelander C, Lind E and Samuelson L 2010 *Proc IEEE* **98** 2047
[4] Tenne R and Seifert G 2009 *Annu Rev Mater Res* 39 387
[5] Xia Y N, Yang P D, Sun Y G, Wu Y Y, Mayers B, Gates B, Yin Y D, Kim F and Yan Y Q 2003 *Adv Mater* **15** 353
[6] Yang P, Yan R and Fardy M 2010 *Nano Lett* **10** 1529
[7] Tenne R and Rao C N R 2004 *Philos T Roy Soc* A **362** 2099
[8] Jamnik J, Dominko R, Erjavec B, Remskar M, Pintar A and Gaberscek M 2009 *Adv Mater* **21** 2715
[9] Mieszawska A J, Jalilian R, Sumanasekera G U and Zamborini F P 2007 *Small* **3** 722
[10] Ruda H, Polanyi J, Yang J, Wu Z, Philipose U, Xu T, Yang S, Kavanagh K, Liu J, Yang L, Wang Y, Robbie K, Yang J, Kaminska K, Cooke D, Hegmann F, Budz A and Haugen H 2006 *Nanoscale Res Lett* **1** 99
[11] Hernandez-Uresti D B, Sánchez-Martínez D, Martínez-de la Cruz A, Sepúlveda-Guzmán S and Torres-Martínez L M 2014 *Ceram Int* **40** 4767
[12] Zanetti S M, Rocha K O, Rodrigues J A J and Longo E 2014 *Sens Actuat* B **190** 40
[13] Anandan S and Wu J J 2014 *Ultrason Sonochem* **21** 1284
[14] Ramos-Delgado N A, Hinojosa-Reyes L, Guzman-Mar I L, Gracia-Pinilla M A and Hernández-Ramírez A 2013 *Catal Today* **209** 35
[15] Han X, Han X, Li L and Wang C 2012 *New J Chem* **36** 2205
[16] Bai S, Zhang K, Luo R, Li D, Chen A and Liu C C 2012 *J Mater Chem* **22** 12643
[17] Arutanti O, Ogi T, Nandiyanto A B D, Iskandar F and Okuyama K 2014 *AIChE Journal* **60** 41
[18] Luo X, Liu F, Li X, Gao H and Liu G 2013 *Mater Sci Semicon Proc* **16** 1613
[19] Amano F, Ishinaga E and Yamakata A 2013 *J Phys Chem C* **117** 22584
[20] Ritala M and Leskelä M 2002 *Chapter 2 - Atomic layer deposition* in Handbook of Thin Films (Academic Press) 103
[21] Available from: www.aldpulse.com.
[22] Johnson R W, Hultqvist A and Bent S F 2014 *Mater Today* **17** 236
[23] Parsons G N, George S M and Knez M 2011 *MRS Bull* **36** 865
[24] Niinisto L, Paivasaari J, Niinisto J, Putkonen M and Nieminen M 2004 *Phys Status Solidi* A **201** 1443
[25] Cheol Seong Hwang C Y Y 2014 *Atomic Layer Deposition for Semiconductors*: Springer New York Heidelberg Dordrecht London)
[26] Pinna N and Knez M 2011 *Atomic Layer Deposition of Nanostructured Materials* (Weinheim, Germany: Wiley-VCH Verlag & Co. KGaA)
[27] Robertson J 2004 *Eur Phys J-Appl Phys* **28** 265
[28] Leskelä M and Ritala M *J Solid State Chem* **171** 170
[29] George S M 2009 *ChemRev* **110** 111
[30] Puurunen R L 2005 *J Appl Phys* **97** 121301
[31] Zaera F 2013 *Coordin Chem Rev* **257** 3177
[32] Putkonen M, Sajavaara T, Niinisto L and Keinonen J 2005 *Anal Bioanal Chem* **382** 1791
[33] Devi A 2013 *Coordin Chem Rev* **257** 3332
[34] Emslie D J H, Chadha P and Price J S 2013 *Coordin Chem Rev* **257** 3282
[35] Suntola T 1996 *Appl Surf Sci* **100** 391
[36] Niinisto L, Ritala M and Leskela M 1996 *Mat Sci Eng B* **41** 23
[37] Elers K E, Blomberg T, Peussa M, Aitchison B, Haukka S and Marcus S 2006 *ChemVapor Dep* **12** 13
[38] Ritala M 1997 *Appl Surf Sci* **112** 223
[39] Mikko R and Markku L 1999 *Nanotechnology* **10** 19
[40] Miikkulainen V, Leskela M, Ritala M and Puurunen R L 2013 *J Appl Phys* **113** 021301

[41] Lee S W, Choi B J, Eom T, Han J H, Kim S K, Song S J, Lee W and Hwang C S 2013 *Coordin Chem Rev* **257** 3154
[42] Sundberg P and Karppinen M 2014 *Beilstein J Nanotechn* **5** 1104
[43] Hatanpää T, Ritala M and Leskelä M 2013 *Coordin Chem Rev* **257** 3297
[44] Knisley T J, Kalutarage L C and Winter C H 2013 *Coordin Chem Rev* **257** 3222
[45] Ramos K B, Saly M J and Chabal Y J 2013 *Coordin Chem Rev* **257** 3271
[46] Tynell T and Karppinen M 2014 *Semicond Sci Tech* **29** 043001
[47] Putkonen M and Niinistö L 2005 *Organometallic Precursors for Atomic Layer Deposition* in Precursor Chemistry of Advanced Materials (Springer Berlin Heidelberg) 125 p
[48] Parsons G N, Atanasov S E, Dandley E C, Devine C K, Gong B, Jur J S, Lee K, Oldham C J, Peng Q, Spagnola J C and Williams P S 2013 *Coordin Chem Rev* **257** 3323
[49] Ponraj J S, Attolini G and Bosi M 2013 *Crit Rev Solid State* **38** 203
[50] Kim H 2003 *J Vac Sci Technol B* **21** 2231
[51] Knoops H C M, Langereis E, van de Sanden M C M and Kessels W M M 2010 *J Electrochem Soc* **157** G241
[52] Potts S E and Kessels W M M 2013 *Coordin Chem Rev* **257** 3254
[53] Mackus A J M, Bol A A and Kessels W M M 2014 *Nanoscale* **6** 10941
[54] Leskelä M and Ritala M 2002 *Thin Solid Films* **409** 138
[55] Knez M, Nielsch K and Niinistö L 2007 *AdvMater* **19** 3425
[56] Detavernier C, Dendooven J, Sree S P, Ludwig K F and Martens J A 2011 *Chem Soc Rev* **40** 5242
[57] Kim H, Lee H B R and Maeng W J 2009 *Thin Solid Films* **517** 2563
[58] Liu M N, Li X L, Karuturi S K, Tok A I Y and Fan H J 2012 *Nanoscale* **4** 1522
[59] Bakke J R, Pickrahn K L, Brennan T P and Bent S F 2011 *Nanoscale* **3** 3482
[60] Marichy C and Pinna N 2013 *Coordin Chem Rev* **257** 3232
[61] Santala E, Kemell M, Leskela M and Ritala M 2009 *Nanotechnology* **20** 035602
[62] Peng Q, Sun X Y, Spagnola J C, Hyde G K, Spontak R J and Parsons G N 2007 *Nano Lett* **7** 719
[63] Heikkila P, Hirvikorpi T, Hilden H, Sievanen J, Hyvarinen L, Harlin A and Vaha-Nissi M 2012 *J Mater Sci* **47** 3607
[64] Peng Q, Sun X-Y, Spagnola J C, Saquing C, Khan S A, Spontak R J and Parsons G N 2009 *ACS Nano* **3** 546
[65] Kim G-M, Lee S-M, Knez M and Simon P 2014 *Thin Solid Films* **562** 291
[66] Cho S, Kim D H, Lee B S, Jung J, Yu W R, Hong S H and Lee S 2012 *Sensor Actuat B* **162** 300
[67] Kim W S, Lee B S, Kim D H, Kim H C, Yu W R and Hong S H 2010 *Nanotechnology* **21** 245605
[68] Lee B S, Kim W S, Kim D H, Kim H C, Hong S H and Yu W R 2011 *Smart Mater Struct* **20** 105019
[69] Kim G M, Lee S M, Michler G H, Roggendorf H, Gosele U and Knez M 2008 *Chem Mater* **20** 3085
[70] Jung M-H, Yun Y J, Chu M-J and Kang M G 2013 *Chem Eur J* **19** 8543
[71] Jeun J H, Park K Y, Kim D H, Kim W S, Kim H C, Lee B S, Kim H, Yu W R, Kang K and Hong S H 2013 *Nanoscale* **5** 8480
[72] Donmez I, Kayaci F, Ozgit-Akgun C, Uyar T and Biyikli N 2013 *J Alloy Comp* **559** 146
[73] Oldham C J, Gong B, Spagnola J C, Jur J S, Senecal K J, Godfrey T A and Parsons G N 2010 *ECS Transactions* **33** 279
[74] Oldham C J, Gong B, Spagnola J C, Jur J S, Senecal K J, Godfrey T A and Parsons G N 2011 *J Electrochem Soc* **158** D549
[75] Kayaci F, Ozgit-Akgun C, Donmez I, Biyikli N and Uyar T 2012 *ACS Appl Mater Interface* **4** 6185
[76] Kayaci F, Ozgit-Akgun C, Biyikli N and Uyar T 2013 *RSC Adv* **3** 6817

[77] Kayaci F, Vempati S, Donmez I, Biyikli N and Uyar T 2014 *Nanoscale* **6** 10224
[78] Kayaci F, Vempati S, Ozgit-Akgun C, Biyikli N and Uyar T 2014 *Appl Catal B* **156** 173
[79] McClure C D, Oldham C J, Walls H J and Parsons G N 2013 *J Vac Sci Technol A* **31** 061506
[80] McClure J P, Devine C K, Jiang R Z, Chu D, Cuomo J J, Parsons G N and Fedkiw P S 2013 *J Electrochem Soc* **160** F769
[81] Li Y, Sun Y J, Xu G J, Lu Y, Zhang S, Xue L G, Jur J S and Zhang X W 2014 *J Mater Chem A* **2** 11417
[82] Szilágyi I, Santala E, Heikkilä M, Kemell M, Nikitin T, Khriachtchev L, Räsänen M, Ritala M and Leskelä M 2011 *J Therm Anal Calorim* **105** 73
[83] Szilágyi I M, Santala E, Heikkilä M, Pore V, Kemell M, Nikitin T, Teucher G, Firkala T, Khriachtchev L, Räsänen M, Ritala M and Leskelä M 2013 *Chem Vapor Dep* **19** 149
[84] Du Q, Wu J B and Yang H 2014 *ACS Catal* **4** 144
[85] Kayaci F, Vempati S, Ozgit-Akgun C, Donmez I, Biyikli N and Uyar T 2014 *Nanoscale* **6** 5735

Journal of Physics: Conference Series **559** (2014) 012011 doi:10.1088/1742-6596/559/1/012011

Fine control of the amount of preferential <001> orientation in DC magnetron sputtered nanocrystalline TiO_2 films

B Stefanov, C G Granqvist and L Österlund

The Ångström Laboratory, Department of Engineering Sciences, Uppsala University, P.O. Box 534, SE-75121 Uppsala, Sweden

E-mail: lars.osterlund@angstrom.uu.se

Abstract. Different crystal facets of anatase $TiO_2$are known to have different chemical reactivity; in particular the {001} facets which truncates the bi-tetrahedral anatase morphology are reported to be more reactive than the usually dominant {101} facets. Anatase TiO_2thin films were deposited by reactive DC magnetron sputtering in Ar/O_2 atmosphere and were characterized using Rietveld refined grazing incidence X-ray diffraction, atomic force microscopy and UV/Vis spectroscopy. By varying the partial O2 pressure in the deposition chamber, the degree of orientation of the grains in the film could be systematically varied with preferred <001> orientation changing from random upto 39% as determined by March-Dollase method. The orientation of the films is shown to correlate with their reactivity, as measured by photo-degradation of methylene blue in water solutions. The results have implications for fabrication of purposefully chemically reactive thin TiO_2 films prepared by sputtering methods.

1. Introduction

Nanocrystalline TiO_2 films have beneficial physicochemical and optical properties in a number of applications related to environmental photocatalysis and energy conversion technologies [1]. Normally anatase particles are dominated by the low energy {101} surface planes with only a fraction of {001} facets, which gives the particles their characteristic truncated bi-pyramidal morphology.

Thin TiO_2 films prepared by reactive DC magnetron sputtering in O_2 atmosphere are initially amorphous, but after heat-treatment it has been reported that preferential alignment of grains occurs in the <004> direction, which exposes reactive (001) planes at the film surface [2,3]. This is desirable, since it is known that the reactivity of anatase depends on the surface structure. For example increased orientation in the <001> direction can favor water splitting reactions, or improve the photonic efficiency of anatase photocatalyst [4-6].

In the present work we demonstrate the possibility to purposefully control the fraction of preferred <001> orientation up to approximately 39%, as determined by March-Dollase analysis of x-ray diffraction data, by adjustment of the partial oxygen pressure for the deposition process. The fraction of preferred orientation was estimated using GIXRD and the effect of the deposition parameters on the crystalline and optical properties of the films was reviewed.

2. Deposition and characterization of anatase TiO_2 films

Standard microscope glass slides were used as substrates. The deposition was done using a pair of 5 cm diameter Ti targets (99.995% purity) positioned symmetrically at a distance of 13 cm from the substrate in a setup described elsewhere [7]. Sputtering was carried out in Ar/O_2 atmosphere (99.998%

purity gases) at a total pressure of 20 mTorr and with O_2 partial pressures in the range 0.65 – 1.54 mTorr (refer to table 1). A constant current of 0.75 A was employed leading to sputtering power in the range of 206 – 280 W with increasing oxygen flow. After deposition the films were heat-treated for 1 h at 500°C.

The phase composition of the films was investigated by use of grazing incidence X-ray diffraction (GIXRD), employing a Siemens D5000 diffractometer using CuKα1 radiation (1.5406 Å) at 0.5° incidence angle.

Additional information about surface morphology was acquired through atomic force microscopy (AFM) employing a PSIA XE150 SPM/AFM instrument.

Film thicknesses were determined by surface profilometry (Dektak XT Advance from Bruker).

A Perkin-Elmer Lambda 900 spectrophotometer equipped with a $BaSO_4$ coated integrating sphere was used for UV-VIS transmittance and reflectance measurements in the range 300 – 800 nm.

Table 1. Summary of the deposition conditions and structural and optical properties of anatase films, deposited at different partial O_2 pressures. Oxygen partial pressure during sputtering (P_{O_2}), film thickness (d), film deposition rate (r_{dep}), fraction of <001> grain orientation ($\eta_{<hkl>}$), refractive index (n), porosity, surface roughness (R_{rms}), and optical bandgap (E_g).

P_{O2} mTorr	d nm	r_{dep} nm min^{-1}	$\eta_{<001>}$ %	n	Porosity %	R_{rms} nm	E_g eV
0.65	574	16	1.99	2.23	88	1.3	3.31
0.95	578	17	12.1	2.15	84	1.1	3.29
1.25	664	18	25.4	2.04	79	1.4	3.28
1.54	697	20	38.6	1.96	75	1.6	3.27

3. Results and discussion

Film deposition was performed at a rate between 16 and 20 nm min^{-1} (proportional to sputtering power) yielding final thicknesses between 574 to 696 nm, as determined both by profilometry and analysis of UV-VIS transmittance data.

Rietveld refinement of the GIXRD data was done using the PowderCell package [8], and the March-Dollase (MD) method was used to determine the fraction of <001> preferred orientation by means of the Zolotoyabko equation [9]:

$$\eta_{\langle hkl\rangle} = \frac{(1-r_{\langle hkl\rangle})^3}{(1-r_{\langle hkl\rangle})^3} \times 100\% \qquad (1)$$

The MD method assumes that the sample has cylindrical orientation symmetry (i.e. the orientation vector does not change by azimuthal rotation, which is applicable in our case since the samples were deposited using a rotating holder). The MD parameter $\eta_{\langle hkl\rangle}$ is a measure of the fraction of crystallites oriented with the crystallographic direction of interest towards the surface (see figure 1).

Analysis of the XRD data shows that $\eta_{\langle 001\rangle}$ increases with increasing O_2 partial sputtering pressure P_{O2} (as can be seen from figure 2 and table 1). It is also evident from figure 2 that the as-deposited films (lowest diffractogram) are amorphous.

The inset in figure 2 also shows photography images of the as-deposited films. The color of the films reveals that films deposited at low P_{O2} are under-stoichiometric. All post-heat treated films were transparent. It has been reported that sputter deposited TiO_2 is usually <001> orientated [3].

Journal of Physics: Conference Series **559** (2014) 012011 doi:10.1088/1742-6596/559/1/012011

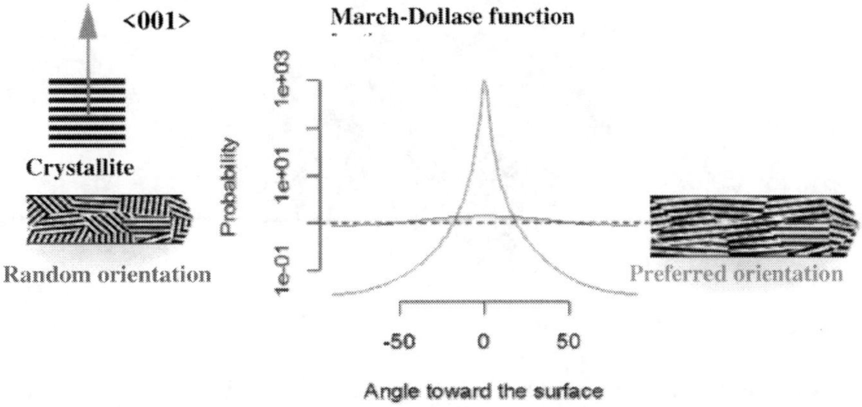

Figure 1. The March-Dollase function for crystallite orientation distribution. The insets represent two examples of films with random and preferred orientation.

We suggest that the introduction of non-stoichiometric growth sites induces crystallites with random growth leading to lower total amount of preferred orientation in the samples. The mean crystallite sizes were estimated to be between 25 and 18 nm, and inversely proportional to P_{O2}.

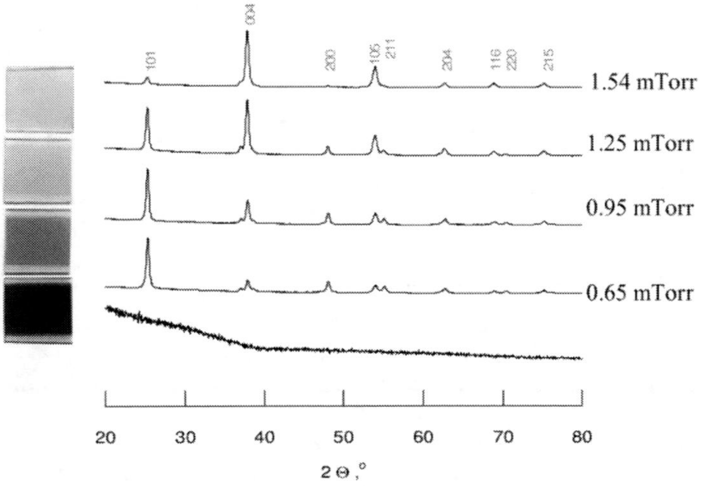

Figure 2. GI-XRD diffractograms for the post-heat treated samples, deposited at different P_{O2} pressures. The insets on the left represent the color of the as-deposited films, which were amorphous as visible by the lowermost diffractogram..

Analysis of AFM data shows that all films were equally smooth and dense regardless of P_{O2} with spherical surface features with sizes comparable with the mean crystalline size (figure 3). The root mean square roughness of the films was about 2.5 nm irrespectively of the sputtering conditions.

doi:10.1088/1742-6596/559/1/012011

Figure 3. Example of the surface topology of a sputtered anatase TiO_2 film. R_{rms} was found to be about 2.5 nm regardless of sputtering conditions.

The optical properties of the films including refractive index and optical bandgap were determined from UV-VIS spectrophotometry. The refractive index, estimated using the Swanepoel approach [10], decreased with increasing P_{O2} from 2.23 to 1.96 (literature value is 2.56, see figure 4). The packing density was determined from the Pulker equation and it was found to be correlated to an increased porosity at higher P_{O2} of about 10% (which is not detectable by AFM). The optical bandgap was estimated according to the method suggested by Hong et al. [11] revealing an approximately constant bandgap with the experimental measurement error (see table 1).

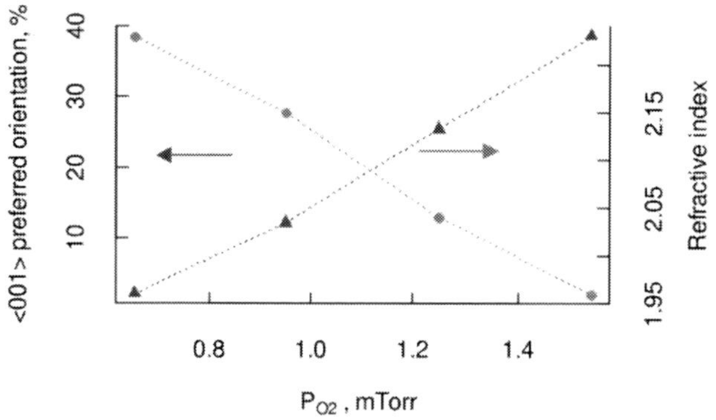

Figure 4. Dependence of <001> preferred orientation fraction and the refractive index of the samples by the oxygen partial pressure.

Table 2. Rate constant for methylene blue photodegradation for TiO_2 films with different $\eta_{<001>}$ [12].

P_{O2}, mTorr	$\eta_{<001>}$, %	r, min^{-1}
0.65	1.99	0.95
0.95	12.1	0.99
1.25	25.4	1.43
1.54	38.6	2.49

Table 2 shows the results from photodegradation of methylene blue for the various TiO_2 thin films used in this study, as described in detail elsewhere [12]. It is evident that the photodegradation rate increases as $\eta_{\langle 001 \rangle}$ increases, and that an approximately 262% increase of the rate constant is obtained at $\eta_{\langle 001 \rangle} = 39$ % compared with an almost randomly oriented film with $\eta_{\langle 001 \rangle}$ **=2 %**.

4. Conclusions

We have presented an approach to fine-tune and quantify the fraction of <001> oriented crystallites, $\eta_{<001>}$, in DC magnetron sputtered anatase TiO_2 films by means of controlling the partial O_2 pressure in the deposition atmosphere, P_{O2}. It was found that increasing P_{O2} leads to increased $\eta_{<001>}$, and porosity, and decreased n. Measurements of the photocatalytic degradation of methylene blue in aqueous solution show that $\eta_{<001>}$ scales with the degradation rate, and hence that photocatalytic reactive TiO_2 thin films can be prepared by purposefully increasing $\eta_{<001>}$ by adjusting sputtering conditions. Such films with controlled fraction of <001> orientation and optical properties might find applications in photocatalysis, as self-cleaning surfaces, or as optical coatings.

5. Acknowledgements

This work was funded by the European Research Council under the European Community's Seventh Framework Program (FP7/2007-2013)/ERC Grant Agreement No. 267234 ("GRINDOOR").

The paper was presented at the INERA Workshop "Transition Metal Oxide Thin Films: Functional Layers in 'Smart Windows' and Water Splitting Devices", September 4–6th, 2014, Varna, Bulgaria. The Workshop is part of the Program of the INERA REGPOT Project of the Institute of Solid State Physics, Bulgarian Academy of Sciences.

References

[1] Hagfeldt A and Graetzel M 1995 *Chem. Rev.* **95** 49-68
[2] Sério S, Melo Jorge M E, Maneira M J P and Nunes Y 2011 *Mater. Chem Phys* **126** 73-81
[3] Howitt D G and Harker A B 1987 *J. Mater. Res* **2** 201-210
[4] Hengerer R, Kavan L, Krtil P and Graetzel M 2000 *J. Electrochem. Soc.* **147** 1467
[5] Deng H, Zhang H and Lu Z 2002 *Chem. Phys. Lett.* **363** 509-514
[6] Zheng J-Y, Bao S-H, Guo Y and Jin P 2014 *ACS Appl. Mater. Interfaces* **6** 5940-5946
[7] Kharrazi M, Azens A, Kullman L and Granqvist C G 1997 *Thin Solid Films* **295** 117-121
[8] Kraus W and Nolze G 1996 *J. Appl. Crystallograph.* **29** 301-303
[9] Dollase W A 1986 *J. Appl. Crystallograph.* **19** 267-272
[10] Swanepoel R 1983 *J. Phys. E Sci. Instrum.* **16** 1214-1222
[11] Hong S, Kim E, Kim D.-W., Sung T.-H. and No K 1997 *J. Non-Cryst. Solids* **221** 245-254
[12] Stefanov B and Österlund L 2014 *Coatings* **4** 587-601

Porous Nickel Oxide Film Sensor for Formaldehyde

U Cindemir, Z Topalian, L Österlund, C G Granqvist and G A Niklasson
Department of Engineering Sciences, The Ångström Laboratory, Uppsala University, Uppsala, Sweden

E-mail: umut.cindemir@angstrom.uu.se

Abstract. Formaldehyde is a volatile organic compound and a harmful indoor pollutant contributing to the "sick building syndrome". We used advanced gas deposition to fabricate highly porous nickel oxide (NiO) thin films for formaldehyde sensing. The films were deposited on Al_2O_3 substrates with prefabricated comb-structured electrodes and a resistive heater at the opposite face. The morphology and structure of the films were investigated with scanning electron microscopy and X-ray diffraction. Porosity was determined by nitrogen adsorption isotherms with the Brunauer-Emmett-Teller method. Gas sensing measurements were performed to demonstrate the resistive response of the sensors with respect to different concentrations of formaldehyde at 150 °C.

1. Introduction

The development of low-cost, sensitive, selective and reliable sensors for detecting volatile organic compounds (VOCs) is gaining importance, and there is great need for monitoring environmental health risks. It is important to quantify the concentration of aldehyde pollutants in indoor environments, since it is well documented that they can have adverse health effects. For example, acetaldehyde and formaldehyde can be generated from building materials such as laminates, furniture, wood ceilings, emulsion paints, etc. The well-known effect of VOCs is the "sick building syndrome", a term used to describe situations where in building inhabitants develop diffuse disease symptoms and discomfort connected with time spent inside the building but without acquiring any specific illness [1].Furthermore high concentrations (>50 ppm) of VOCs are extremely irritating and probably carcinogenic [2]. Since people in industrialized countries spend 80 to 90% of their time indoors, in buildings or vehicles, the monitoring of these compounds is of large significance [3].

Nanostructured metal oxides can exhibit semiconducting properties and are attractive as low-cost chemical detectors. Nanocrystalline nickel oxide is of special interest for gas sensing at low operating temperatures [4]. The mechanism for gas detection relies on a change of resistance due to gas adsorption on the surface of a thin film. Different production techniques, such as sputtering and evaporation are being used to manufacture nano-crystalline porous thin film for gas sensing purposes [5–7]. However, those techniques normally yield a rather uncontrolled, broad distribution of particle sizes that may be disadvantageous for sensing applications. However, a technique called Advanced Gas Deposition (AGD) allows the production of thin films with well-controlled, log-normal particle size distributions [8–11]. AGD uses thermal evaporation and comprises two chambers connected via a pipe. The first chamber where the nanoparticles are produced is called the evaporation chamber and has a higher pressure than in the second chamber (referred as the deposition chamber), where the film is deposited onto the substrate by a spray.

2. Materials and methods

2.1. Synthesis of NiO nanoparticles and thin film sensors

The pressures in the AGD chambers to produce nanoparticle films were set to 86.5 mbar and 8.45 mbar for the evaporation and deposition chamber, respectively. During the process, Ni metal was melted and vaporized by use of 2.5 kW of RF power. To produce NiO during the process, 20 l/min of He and 100 ml/min of O_2 were injected into the evaporation chamber. NiO nanoparticles were deposited onto glass and also onto silicon substrates in order to allow material characterization. To manufacture sensor devices, the films were deposited onto aluminium oxide (Al_2O_3) substrates with prefabricated comb-structured platinum–silver alloy (C4729) electrodes (0.3 mm spacing) on one side and a resistive heater at the opposite side. The dimensions of the ceramic substrates were 25×5×0.5 mm. The thickness of films on single crystal Si substrates was 1.5 μm, measured by a profilometer (Bruker DektakXT). Thicker films are needed to facilitate accurate resistance measurements. However, the decreased resistance is related to decreasing the film porosity and a possible lowering of sensor performance.

2.2. X-ray diffraction (XRD)

For XRD measurements, a glass substrate coated with NiO nanoparticles was loaded into a Philips D5000 diffractometer equipped to detect with CuK_α (λ=1.54 Å) radiation. A continuous scan was performed in the 30–80° range of diffraction angles (2θ) using an integration step width of 0.05°. The measuring time per step was 12 seconds. Structural parameters were extracted from θ–2θ scans, and Scherrer's formula [12] was used to calculate average crystallite size.

2.3. Scanning electron microscopy(SEM)

Samples for SEM analysis were deposited on Si wafer pieces. SEM images were acquired using a Zeiss LEO1550 instrument with InLens detector and a working distance of 2.7 mm. Due to the low electrical conductivity of NiO, the acceleration voltage was kept at 5 kV thereby limiting the resolution and magnification.

2.4. Brunauer-Emmett-Teller (BET) measurements of surface area

The specific surface area was determined with the BET method using NiO powder collected from thick layers of NiO nanoparticles coated onto the glass substrates and then the powder was removed mechanically. BET isotherm measurements were done using nitrogen gas as adsorbing agent, and the temperature was set to -195.8°C. The sample mass employed, after degassing, was measured to be 122.4 mg.

2.5. X-ray photo-electron spectroscopy (XPS)

Samples for XPS measurements were coated on Si wafer substrates. In the XPS spectra, oxygen 1s features (O1s) were analysed in order to determine the chemical composition of the nanoparticles. XPS analyses were performed using a PHI Quantum 2000 instrument, equipped with an Al x-ray source having an energy of 1486.7 eV. The carbon peak at 284.3 eV due to residual aliphatic carbon was used as a reference for the energy scale.

2.6. Gas sensing measurements

The gas sensing setup consisted of a 300 ml cylindrical chamber, connected to a gas handling system via mass flow controllers. The sensor's resistance response to formaldehyde was recorded at a substrate temperature of 150°C and dry synthetic air (80% N_2 and 20% O_2) was used as carrier gas. Three cycles of exposure to formaldehyde (30 min) and synthetic air (30 min) were performed. The sensors were biased with a battery and a 200 MΩ load resistor, and sensor resistances were extracted from the voltage on the sensor and the load resistor.

Journal of Physics: Conference Series **559** (2014) 012012 doi:10.1088/1742-6596/559/1/012012

3. Results and discussion

3.1. Material characterization

The porosity and the particle size of the NiO films were characterized via SEM analysis. The SEM image in figure 1 shows a highly porous surface comprised of particles with sizes less than 10 nm together with voids among the particles.

XRD analysis revealed a film structure consistent with the cubic phase (fcc) of NiO (JCPDS–ICDD No. 47–1049), with a cell parameter of 4.17 Å (figure 2).The crystallite size was calculated by Scherrer's formula and found to be approximately 3.7 nm. .The specific surface area of the NiO particles deposits was determined to be 184 m^2/g. The average pore size was found to be 4.3 nm, using the Barrett-Joyner-Halenda method [13]. XPS data of the O1s feature showed two peaks: a predominant one due to NiO located at ~530 eV binding energy and a minor contribution of Ni_2O_3, around 532 eV binding energy.

Figure 1. SEM image of NiO film showing a porous surface with particles sizes less than 10 nm.

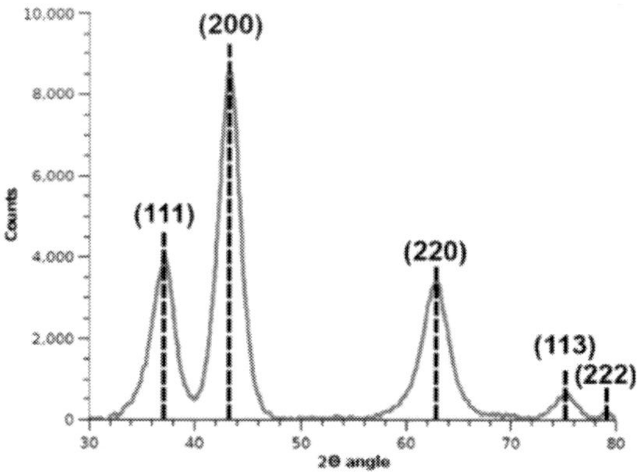

Figure 2. XRD spectra of a NiO film with (*hkl*) indices pertinent to the fcc NiO structure.

3.2. Gas sensing measurements

Resistance changes were employed as the sensing mechanism for formaldehyde detection in the NiO sensor devices. Gas sensing measurements were done at a total gas flow rate of 200 ml/min. Sensor resistances were recorded with formaldehyde diluted in synthetic air at formaldehyde concentrations of 50, 25, 10 and 5 ppm. The underlying mechanism for the resistance change in the NiO film is mainly governed by oxygen adsorption in the film and ensuing electron transfer reactions following oxidation of formaldehyde at the NiO surface [14]. Oxygen is adsorbed on the surface of NiO (*p*-type semiconductor) and dissociates into atomic O fragments, as explained elsewhere [14]. Bonding of O fragments results in the formation of Ni^{3+}centres which increase the concentration of charge carriers (holes). Formaldehyde reacts with O adatoms which results in reduction of Ni^{3+} sites, and hence in a decreasing concentration of hole carriers (h^+) in the NiO film, and in an increased film resistance. The relevant chemical reactions at the NiO surface were proposed to be [5]:

$$Ni^{2+} + O_{(ads)} \rightarrow Ni^{3+} + O^{-}_{(ads)}, \quad (1)$$

$$HCHO_{(gas)} + O^{-}_{(ads)} + h^{+} \rightarrow HCOOH_{(gas)}, \quad (2)$$

$$HCHO_{(gas)} + 2O^{2-}_{(ads)} + 4h^{+} \rightarrow CO_{2(gas)} + H_2O_{(gas)}, \quad (3)$$

The sensitivity S can be defined as the change in resistance upon exposure to formaldehyde, divided by the initial resistance, according to

$$S = \frac{R_{gas} - R_{air}}{R_{air}} \times 100\% \qquad (4)$$

where R_{gas} and R_{air} are the resistance after exposure to formaldehyde and synthetic air, respectively. An average was taken over three measurement cycles. The resistance drift is defined as the ratio of the time-averaged resistance in synthetic air, after and before the three cycles of gas exposure, *i.e.*

$$R_{drift} = \frac{R_{air_{start}} - R_{air_{end}}}{R_{air_{end}}} \times 100\% \qquad (5)$$

Here R_{air_start} stands for time-averaged resistance of the sensor in synthetic air before exposure to formaldehyde for three cycles, whereas R_{air_end} is the corresponding measure after exposure.

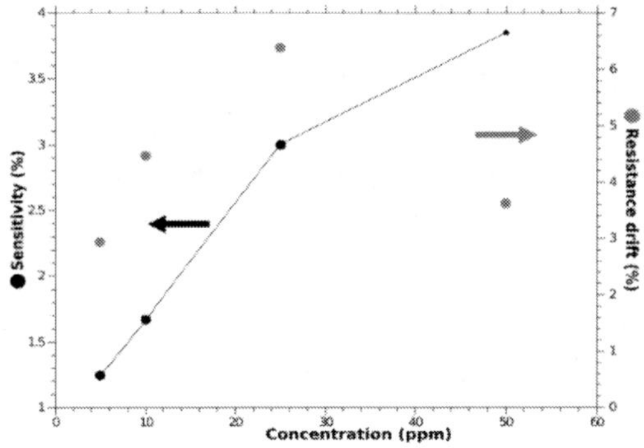

Figure 3. Dependence of Sensitivity, S, and resistance drift, R_{drift}, of a NiO sensor exposed to different concentrations of formaldehyde.

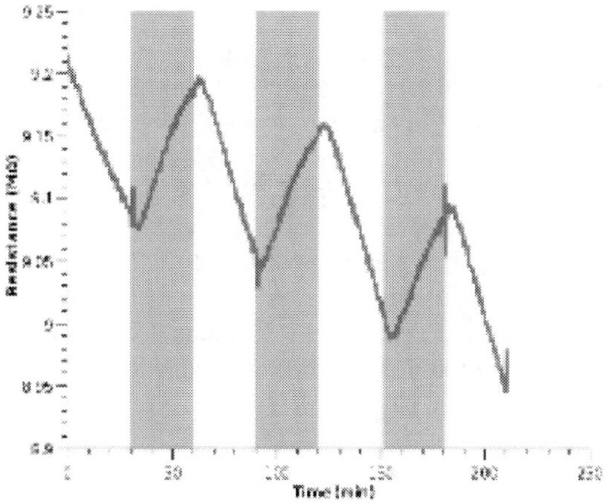

Figure 4. Transient resistance response of a NiO sensor at 150°C when exposed to 5 ppm of formaldehyde with synthetic air as a carrier gas. Shaded regions refer to inlet of formaldehyde.

The resistivity measurements yielded a turn-on time between gas exposure and sensing of 2.5 minutes and a saturation time (time to reach 90% of the resistance change before the end of exposure)of around 20 minutes at 50 ppm. The NiO films showed sensitivity values for formaldehyde detection in the range from 1.2 to 3.8% for 5 and 50 ppm of formaldehyde, respectively (figure 3). The resistance drift, as defined in equation 5, was measured to be between 2.9 and 6.4% depending on gas concentrations (figure 4).

4. Conclusion

It is shown that nanoporous NiO particle films produced by AGD are sensitive to low levels of formaldehyde (of the order 1 ppm) at a sensor temperature of 150°C. The results are promising for further development of inexpensive indoor gas sensing devices. It would be desirable to further increase the sensitivity, as well as to decrease the operating temperature and the response time.

Acknowledgements
This work has been supported by the European Research Council under the European Community's Seventh Framework Program (FP7/2007–2013)/ERC Grant Agreement No. 267234 (GRINDOOR).The paper was presented at the INERA Workshop "Transition Metal Oxide Thin Films: Functional Layers in 'Smart Windows' and Water Splitting Devices", September 4–6th, 2014, Varna, Bulgaria. The Workshop was part of the Program of the INERA REGPOT Project of the Institute of Solid State Physics, Bulgarian Academy of Sciences.

References

[1] Murphy M 2006 *Sick Building Syndrome and the Problem of Uncertainty: Environmental Politics, Technoscience and Women Workers* (Duke University Press, Durham, USA)
[2] Giberti A, Carotta M C, Fabbri B, Gherardi S, Guidi V and Malagù C 2012 *Sensors Actuat. B* **174** 402–405
[3] Godish T 2000 *Indoor Environmental Quality* (CRC Press, Boca Raton, USA)
[4] Luyo C, Ionescu, R, Reyes L F, Topalian Z, Estrada W, Llobet E, Granqvist C G and Heszler P 2009 *Sensors Actuat. B* **138** 14–20
[5] Castro-Hurtado I, Herrán J, Mandayo G G and Castaño E 2011 *Thin Solid Films* **520** 947–952
[6] Kemmler J A, Pokhrel S, Birkenstock J, Schowalter M, Rosenauer A, Bârsan N, Weimar U and Mädler L 2012 *Sensors Actuat. B* **161** 740–747
[7] Du Y, Wang W, Li X, Zhao J, Ma J, Liu Y and Lu G 2012 *Mater. Lett.* **68** 168–170
[8] Granqvist C G and Buhrman R A 1976 *J. Appl. Phys.* **47** 2200–2219
[9] Hayashi C, Uyeda R and Tasaki A, editors 1997 *Ultra-Fine Particles: Exploratory Science and Technology* (Noyes, Westwood, USA)
[10] Söderlund L, Kiss L B, Niklasson G A and Granqvist CG 1998 *Phys. Rev. Lett.* **80** 2386–2388
[11] Reyes L F, Saukko S, Hoel A, Lantto V and Granqvist C G 2004 *J. European Ceramic Soc.* **24** 1415–1419
[12] Cullity B D and Stock S R 2001 *Elements of X-ray Diffraction*, 3rd edition (Prentice-Hall, Upper Saddle River, USA)
[13] Barrett E P, Joyner L G and Halenda P P 1951 *J. American Chem. Soc.* **73** 373–380
[14] Kohl D 2001 *J. Phys. D: Appl. Phys.* **34** R125–R149

Gas sensing properties of very thin TiO_2 films prepared by atomic layer deposition (ALD)

S Boyadjiev[1], V Georgieva[2], L Vergov[2], Zs Baji[3], F Gáber[4] and I M Szilágyi[1,4]

[1] MTA-BME Technical Analytical Research Group, Szent Gellért tér 4., Budapest, H-1111, Hungary;
[2] Georgi Nadjakov Institute of Solid State Physics, Bulgarian Academy of Sciences, blvd Tzarigradsko chaussee 72, 1784 Sofia, Bulgaria;
[3] Hungarian Academy of Sciences, Research Centre for Natural Sciences, Institute of Technical Physics and Materials Science, H-1121 Budapest, Konkoly Thege M. út 29-33, Hungary;
[4] Budapest University of Technology and Economics, Department of Inorganic and Analytical Chemistry, Szent Gellért tér 4., Budapest, H-1111, Hungary

E-mail: boiajiev@gmail.com

Abstract. Very thin titanium dioxide (TiO_2) films of less than 10 nm were deposited by atomic layer deposition (ALD) in order to study their gas sensing properties. Applying the quartz crystal microbalance (QCM) method, prototype structures with the TiO_2 ALD deposited thin films were tested for sensitivity to NO_2. Although being very thin, the films were sensitive at room temperature and could register low concentrations as 50-100 ppm. The sorption is fully reversible and the films seem to be capable to detect for long term. These initial results for very thin ALD deposited TiO_2 films give a promising approach for producing gas sensors working at room temperature on a fast, simple and cost-effective technology.

1. Introduction

Nowadays TiO_2 in thin film form is one of the most extensively studied transition metal oxides. TiO_2 as n-type semiconductor is widely studied mainly for photocatalysis and gas sensing [1]. But it is also an excellent material for many solid-state devices including various optical applications [2], solar cells [3], corrosion-protective coating, gate insulator in MOSFETs and many others [1].

The gas sensing properties of TiO_2 thin films have been widely researched. Nevertheless, still there is lack of knowledge about the gas sensing of ultra thin films with thicknesses of a few nanometers.

ALD is a vapor phase deposition technique based on successive, alternating surface controlled and self-limiting reactions to produce highly conformal and uniform thin films with thickness control of sub-nanometer precision and capable of producing thin films of a variety of materials. ALD also offers exceptional thin film conformality on high-aspect ratio structures and tunable film composition. With these advantages, ALD has emerged as a powerful tool for many industrial and research applications [4-11].

A large number of techniques have been developed for environmental monitoring and control of the toxic gases. Semiconductor thin films such as TiO_2, SnO_2, WO_3, MoO_3 and ZnO have been widely investigated for sensing different toxic gases. These devices are usually not suited to make high precision measurements of gas concentrations but to detect the presence of target gases and give a warning if several threshold values are attained [11-19].

The quartz crystal microbalance (QCM) is a well-established tool for monitoring the adsorption of nano amounts of materials and for detecting physical properties of thin layers deposited on the crystal surface [20,21]. Compared with other sensors, the advantages of QCM sensors are the simple technological implementation, capability of operating at room temperature, good sensitivity and working process reversibility, low consumption and easy real-time monitoring. The QCM-based techniques attract particular interest due to their high sensitivity in the nanogram range and capability of detecting very small amounts. The analytical performance of such sensors strongly depends on the properties of the sensing coating deposited on the QCM electrodes.

Prototype QCM gas sensors with several transition metal oxide films were already made by our team and tested for sensitivity to NO_2 and NH_3 in a specially designed chamber. Among others, the sensing behavior of sputtered TiO_2 thin films towards NO_2 were investigated [22-24].

As a follow up of the previous studies, in the present research we focused on the sensing behavior of very thin TiO_2 films produced by atomic layer deposition (ALD). TiO_2 ALD films were grown either on quartz resonators, which were used for the gas sensing tests to NO_2, or on Si substrates, which were used for further characterization of the ALD grown films.

2. Experimental

The films were prepared using ALD deposition performed in a Picosun SUNALE R-100 reactor at 200°C substrate temperature with titanium tetraisopropoxide (TTIP) and H_2O as precursors. The TTIP precursor was heated at 70°C in order to reach sufficient partial pressure of its vapor. Nitrogen was used as carrier gas while the overall pressure in the reactor chamber was ~10 mbar. The precursor pulse times were 0.1 s for both TTIP and H_2O. The purge time was 3 s after the TTIP pulse and 4 s after the H_2O. TiO_2 thin films were deposited in 150 ALD cycles on both quartz resonators and Si wafers.

The film morphology was investigated by SEM and the composition was studied by energy dispersive X-ray analysis (EDX) in a JEOL JSM-5500LV scanning electron microscope. The thickness of the films was measured by UV-Vis reflectometry using an Avaspec-2048 spectrophotometer.

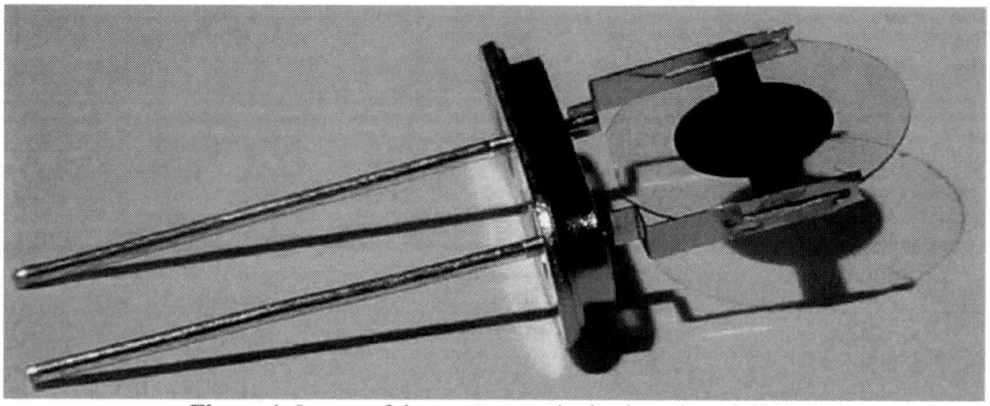

Figure 1. Image of the test sensor device based on QCM.

The sensing properties of the films were studied using quartz resonators and applying the QCM method. The tests were performed with resonators produced on 8-mm polished AT-cut quartz plates with thermally evaporated golden electrodes (diameter 4 mm and thickness of about 120 nm with Cr underlayer) on both sides (figure 1). Their working resonance frequency was around 16 MHz. The initial parameters of the resonator and its quality were evaluated by measuring the equivalent dynamic parameters: static capacitance C_0 and equivalent dynamic resistance Rq using a Selective Level Meter. The dynamic capacitance Cq, the dynamic inductance Lq, and the quality factor Q were obtained by calculation [25]. The sorption of NO_2 was tested in a special home-made measurement system, which was described in details previously [22].

The sorption properties of the TiO_2 films were evaluated by measuring the resonance frequency shift of the QCM structures covered with the TiO_2 thin films at different NO_2 concentrations. The main stages of the measurement consisted of (i) purging of the structures with dry air; (ii) creating a certain concentration of the measured gas in the chamber; (iii) reaching saturation of the frequency values; (iv) purging of the structures with dry air thus restoring the sensor and preparing it for new measurement. The NO_2 concentration in the test chamber was controlled by mass flow controllers (MFCs) for NO_2 and diluting gas flow. The experiments were carried out at constant room temperature. The QCM frequency was registered by frequency counter Hameg 8123 connected to the QCM and to a computer for data recording. The relative error for frequency measurement was $\pm 5.25\text{x}10^{-7}$. The measurements were based on the correlation between the frequency shift and the additional mass loading the resonator. Sauerbrey [26] developed an empirical equation for AT–cut quartz, describing the relation between the mass of the thin film deposited on the quartz crystal and the corresponding change in the resonance frequency, thus allowing the absorbed mass to be calculated. The desorption (recovery) time, t_d, was also measured.

3. Results and discussion

Very thin TiO_2 films with a thickness less than 10 nm were deposited by ALD on both QCMs and Si substrates. With 150 ALD cycles a thickness of ~ 8 nm was achieved, based on the UV-Vis reflectometry measurement of the reference TiO_2 thin films deposited on Si wafers.

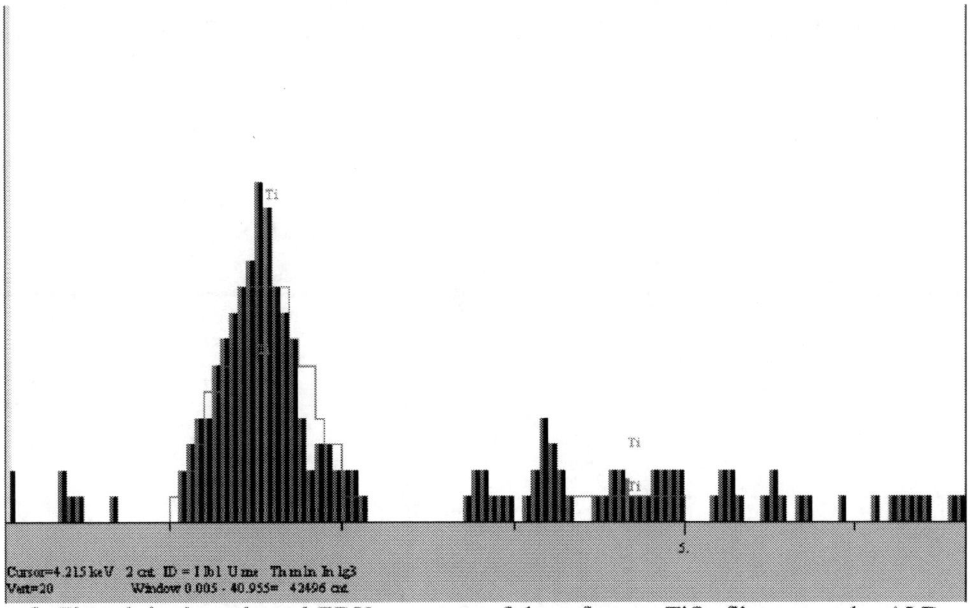

Figure 2. Ti peak in the enlarged EDX spectrum of the reference TiO_2 film grown by ALD on a Si wafer.

After the ALD deposition the film morphology was examined by SEM and their composition was studied by EDX. According to SEM images (not shown here), the films were of good quality, homogeneous, uniform, and without cracks or defects. The successful deposition of TiO_2 was clearly shown by the presence of Ti in the EDX spectrum of the reference TiO_2 film grown by ALD on the Si wafer (figure 2). At this very small TiO_2 film thickness the EDX method had quite big inaccuracy of the oxygen detection, due to receiving signal mainly from the substrate covered by a native oxide layer, and also because adsorbed H_2O could be present on the surface of TiO_2. Thus, the EDX study had only qualitative character and was useful mainly to prove the successful deposition of the very thin TiO_2 films. Results from the EDX study of the reference 8 nm TiO_2 film deposited by ALD on the Si substrate are presented in table 1.

Table 1. EDX results for the ~8 nm thick reference TiO_2 film deposited by ALD on a Si wafer.

Element	Atomic Wt %	Weight %
O	2.68	1.54
Si	97.12	98.22
Ti	0.14	0.24

The gas sensing properties of the TiO_2 films were tested with concentrations in the interval between 10 ppm and 5000 ppm. The films were found to be sensitive even at 50-100 ppm, but at such concentrations their sensitivity was very week and the observed frequency change (Δf) was 3-4 Hz. Typical frequency-time characteristics (FTC) for a well-working sensor were observed at 1000 ppm (figure 3). The response of the sensor was fast (few s), and its sensitivity vs. concentration was close to linear. The recovery was also in a few s timescale. In addition, the recovery was complete and the initial values of the resonator were achieved after removing the NO_2 gas from the chamber. The sorption was considered to be a fully physical process, shown by the full recovery of the sensor for comparably short time.

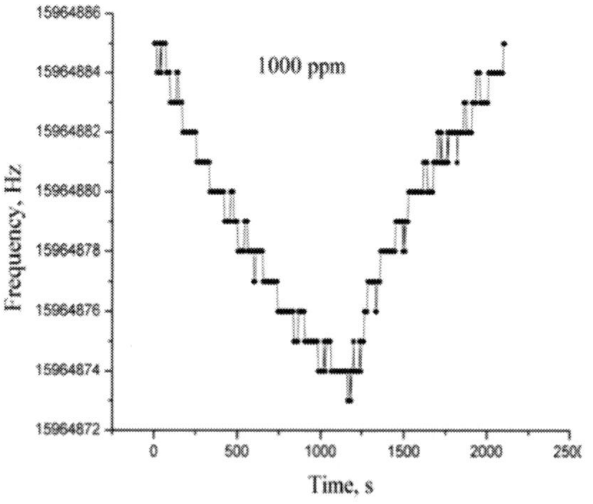

Figure 3. FTC of QCM with ~8 nm ALD deposited TiO_2 thin film during saturation and recovery at NO_2 concentration of 1000 ppm.

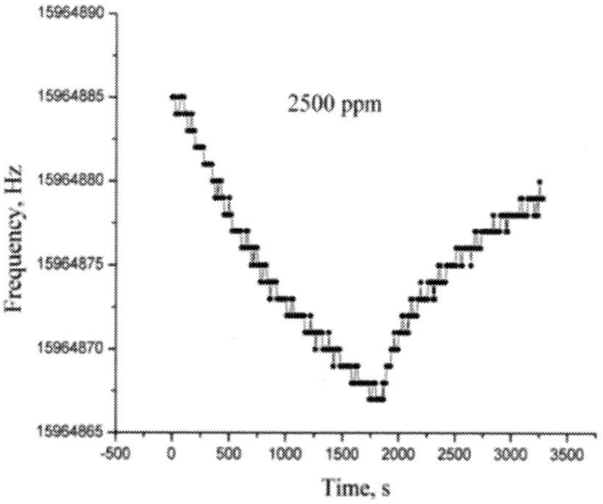

Figure 4. FTC of QCM with ~8 nm ALD deposited TiO_2 thin film during saturation and recovery at NO_2 concentration of 2500 ppm.

At higher concentrations the kinetics of the sorption process was even better expressed and faster. At concentration of 2500 ppm the change of frequency until reaching saturation was $\Delta f = 18$ Hz. The sorption was still physical and the resonator could be fully recovered but the recovery times were longer. The FTC at 2500 ppm is presented in figure 4.

In our previous studies similar results were observed for several times thicker TiO_2 films obtained by reactive sputtering and other methods [22-24]. Here, the results for the sensitivity to NO_2 of the very thin ALD TiO_2 films are very promising. They indicate that even ultra thin films can be implemented in such QCM gas sensors where only the surface properties of the TiO_2 films are used.

The present results are very promising, as they show that even ultra thin ALD TiO_2 films have the required sensitivity for NO_2 sensing. Therefore, these layers can be implemented in QCM gas sensors, because only the surface properties of the films are relevant for such applications.

4. Conclusion

Very thin (~8 nm) TiO_2 films were deposited by ALD and their sensitivity to NO_2 was studied applying the QCM method. Even these very thin ALD TiO_2 films showed good sensitivity to NO_2 at room temperature and capability to register as low concentrations as 50 ppm. The sorption was fully reversible. The ALD TiO_2 films were of good quality, stable and seemed capable for measurements for long terms. The ALD method was found to be suitable for fast and cost-effective deposition of TiO_2 thin films for QCM gas sensor applications.

The main conclusion based on the results obtained is that QCM structures covered with very thin ALD TiO_2 films are suitable for NO_2 detection. These promising results for sensitive films with thickness of less than 10 nm give a hope that, after a further study and development, ultra thin TiO_2 films could be used for gas sensing.

Acknowledgements

This paper was presented at INERA Workshop "Transition Metal Oxide Thin Films-functional Layers in "Smart windows" and Water Splitting devices: Technology and Optoelectronic properties", September 4-6th, 2014, Varna, Bulgaria. The Workshop is part of the Program of INERA REGPOT Project of Institute of Solid State Physics, Bulgarian Academy of Sciences.

S. Boyadjiev thanks the Postdoctoral Fellowship programme of the Hungarian Academy of Sciences (2013-2015).

I. M. Szilágyi acknowledges a János Bolyai Research Fellowship of the Hungarian Academy of Sciences and an OTKA-PD-109129 grant.

References

[1] Diebold U 2003 *Surf Sci Rep* **48** 53
[2] Bennett J M, Pelletier E, Albrand G, Borgogno J P, Lazarides B, Carniglia C K, Schmell R A, Allen T H, Tuttle-Hart T, Guenther K H and Saxer A 1989 *Appl Opt* **28(16)** 3303
[3] Baraton M-I 2011 *Open Nanosci J* **5** 64
[4] Johnson R W, Hultqvist A and Bent S F 2014 *Mater Today* **17(5)** 236
[5] George S M 2009 *Chem Rev* **110** 111
[6] Knez M, Nielsch K and Niinistö L 2007 *Adv Mater* **19** 3425
[7] Niinisto L, Ritala M and Leskela M 1996 *Mat Sci Eng* B **41** 23
[8] Suntola T 1996 *Appl Surf Sci* **100** 391
[9] Szilágyi I M, Santala E, Heikkilä M, Pore V, Kemell M, Nikitin T, Teucher G, Firkala T, Khriachtchev L, Räsänen M, Ritala M and Leskelä M 2013 *Chem Vapor Dep* **19** 149
[10] Szilágyi I M, Teucher G; Härkönen E; Färm E; Hatanpää T; Nikitin T, Khriachtchev L, Räsänen M, Ritala M and Leskelä M 2013 *Nanotechnology* **24** 245701
[11] Lee Y C, Huang H, Tan O K and Tse M S, 2008 *Sens Actuat B* **132** 239
[12] Ali M, Wang C Y, Röhlig C C, Cimalla V, Stauden T and Ambacher O 2008 *Sens Actuat B* **129** 467

[13] Law J B K and Thong J T L 2008 *Nanotechnology* **19** 205502
[14] Szilágyi I M, Saukko S, Mizsei J, Tóth A L, Madarász J and Pokol G 2010 *Solid State Sci* **12** 1857
[15] Szilágyi I M, Wang L, Gouma P I, Balázsi C, Madarász J and Pokol G 2009 *Mater Res Bull* **44** 505
[16] Szilágyi I M, Madarász J, Pokol G, Király P, Tárkányi G, Saukko S, Mizsei J, Tóth A L, Szabó A and Varga-Josepovits K 2008 *Chem Mater* **20** 4116
[17] Balázsi C, Wang L, Zayim E O, Szilágyi I M, Sedlackova K, Pfeifer J, Tóth A L and Gouma P I 2008 *J Eur Ceram Soc* **28** 913
[18] Szilágyi I M, Saukko S, Mizsei J, Király P, Tárkányi G, Tóth A L, Szabó A, Varga-Josepovits K, Madarász J and Pokol G 2008 *Mater Sci Forum* **589** 161
[19] Wang L, Pfeifer J, Balázsi C, Szilágyi I M and Gouma P I 2007 *Proc SPIE* **6769** 67690E
[20] Mirmohseni A and Rostamizadeh K 2006 *Sensors* **6** 324
[21] Cunningham A 1998 *Introduction to Bioanalytical Sensors* (New York: Wiley)
[22] Georgieva V, Stefanov P, Spassov L, Raicheva Z, Atanassov M, Tincheva T, Manolov E and Vergov L 2009 *J. Optoelectr Adv Mater* **11** 1363
[23] Georgieva V, Raicheva Z, Grechnikov A, Gadjanova V, Atanassov M, Lazarov J and Manolov E 2010 *J Phys Conf Ser* **253** 012046
[24] Yordanov R, Boyadjiev S and Georgieva V 2014 *Digest J Nanomat Biostruct* **9** 467
[25] Manolov S and Tihchev H 1982 *Generators* (Sofia: Tehnika)
[26] Sauerbrey G Z 1959 *Physik* **155** 206

Application of electrochemically deposited nanostructured ZnO layers on quartz crystal microbalance for NO_2 detection

B Georgieva[1], M Petrov[2], K Lovchinov[2], M Ganchev[2], V Georgieva[3] and D Dimova-Malinovska[2]

[1]UCTM bul. " St. Kliment Ohridski" № 8, Sofia 1756, Bulgaria
[2]Central Laboratory of Solar Energy and New Energy Sources at the Bulgarian Academy of Sciences, blvd Tsarigradsko chaussee 72, Sofia 1784, Bulgaria
[3]Institute of Solid State Physics at the Bulgarian Academy of Sciences, blvd. Tsarigradsko chaussee 72, Sofia 1784, Bulgaria

E-mail: b.georgiewa@abv.bg

Abstract. The research was fixed on sensing behavior of ZnO nanostructured (NS) films to NO_2 concentrations in the environment. The ZnO NS layers are deposited by electrochemical method on quartz resonators with Au electrodes. The sorption properties of ZnO layers were defined by measuring the resonant frequency shift (Δf) of the QCM-ZnO structure for different NO_2 concentrations. The measurements were based on the correlation between the frequency shift of the QCM and additional mass loading (Δm) on the resonator calculated using Sauerbrey equation for the AT-cut quartz plate. Frequency – Time Characteristics (FTCs) of the samples were measured as a function of different NO_2 concentrations in order to define the sorption abilities of ZnO layers. The experiments were carried out on a special set up in a dynamical regime. From FTCs the response and the recovery times of the QCM-ZnO structure were measured with varying NO_2. Frequency shift changed from 23 Hz to 58Hz when NO_2 was varied in the range of 250ppm – 5000ppm. The process of sorption was estimated as reversible and the sorption as physical. The obtained results demonstrated that QCM covered with the electrochemically deposited nanostructured ZnO films can be used as application in NO_2 sensors.

1. Introduction

Nowadays the atmosphere became polluted caused by industrial development and human activity. Ones of the most dangerous gases are the nitrogen oxides (NO, N_2O, N_2O_2, NO_2). The NOx (combination of NO and NO_2) is a typical oxidizing gas, which is found in the combustion exhaust of automobiles and oil-burning power plants. It is acutely toxic and causes problems such as acid rain, photochemical smog and corrosion. This requires monitoring and controlling these gases and development a wide variety of sensors using different materials and technologies [1, 2].

A quartz crystal microbalance (QCM) is one of extremely sensitive mass devices. It is attractive since King firstly introduced it into analytical chemistry [3]. It can be used as gas sensors [4–6], biosensors [7, 8] etc. The QCM sensor properties as sensitivity, selectivity and time response are strongly influenced by the sensing films properties. These sensors are highly sensitive to mass changes

[1] To whom any correspondence should be addressed.

in the presence of a coating interacting with test gas. The characteristics of QCM gas sensors depend on the materials and the sensing film properties deposited on their electrodes. Many materials have been successfully employed as coatings for QCM sensors [9-11]. ZnO as II–VI semiconductor is one of them. This material posseses high chemical stability, low dielectric constant and high luminous transmittance. As gas sensing material, it is one of the earliest discovered and most widely applied oxide gas sensing material because of its electrcal and chemical properties[12].

ZnO based sensors are sensitive to many sorts of gases, it shows no selectivity for a certain gas. Many attempts are made aimimg to improve its selectivityby using dopants and additives in order to modulate the gas-sensing characteristics [2].

Other approaches for enhancing sensor properties include using the one-dimensional ZnO nanostructures such as nanowires, nanorods, nanobelts and nanotetrapods. They have attracted high attention because of the dimension or surface-to-volume ratio influencing significantlythe material performance [11].

The aim of this study is to investigate NO_2 gas-sensing ability of QCM coated with electrochemically deposited ZnO films towards NO_2 in the air. The sensors characteristics such as response and recovery times, frequency shift and mass-loading of the fabricated structures are studied as a function of different NO_2 concentrations.

2. Experimental

In this work the QCM sensors modified with ZnO nanostructured film are fabricated.

Thin ZnO nanostructured (NS) films are deposited by an electrochemical process from slightly acid aqueous solution of $ZnCl_2$ ($5.\ 10^{-3}$ M) and KCl (0.5 M) with pH 4.0 at 80°C and -1000mV (vs SCE) using a three-electrode electrochemical cell [13]. The electrolyte is agitated by magnetic stirrer. The quartz crystal resonator with Au electrode is used as a cathode. Spectrally pure graphite plate electrode is used as an anode. The deposition is carried out controlling the redox potential of the system. The total oxygen content in the solution is controlled by a DO&T meter Hanna Instruments 9146. The duration of the ZnO deposition is 35 minutes. The study of the structural properties of the NS ZnO layer is performed by SEM microscopy using Philips 515 SEM apparatus.

For gas sensing experiments QCM were fabricated on AT-cut quartz substrates. Quartz plates with 8 mm diameter are used. Au-electrodes 4 mm in diameter are thermally evaporated on both sides with a thickness of 60 nm. The Cr underlayer is used for better adhesion of Au electrodes on the quartz plate. The fabricated 16 MHz QCMs are evaluated by measuring the motional resistance (Rs) and calculating the quality factor (Q). These parameters and the series frequency of the QCM loaded with ZnO films are controlled at each stage of the experiments.

The experimental setup works in a dynamic regime in continuous flow of testing and carrying gases. The sensing properties of the QCM-ZnO to NO_2 are measured by the resonance frequency shift response. The QCM is installed on a special holder inside the test chamber. The temperature of the sample is measured by a Pt-thermosensor placed next to it and is kept constant 24 ± 0.2^0C during the all experiments. The flow rates of the used gases are kept constant by two mass flow controllers (FC-260 and FC-280). Mixing camera provides the homogeneity of the gas mixtures. The ratio of the flows of the test and diluting gases defines NO_2 gas concentration. In the experiments a dry synthetic air – high purity and 10 000 ppm NO_2 diluted in synthetic air are used.

The measurement process consists of three basic stages: purging the camera with air flow until the frequency of the QCM-ZnO reaches a constant value (F_o); switching NO_2 and reaching the saturation frequency value (F); purging the camera with dry air until approaching the initial frequency of the investigated structure. The measurements have been carried out continuously with two seconds interval. A frequency counter Hameg 8123 is connected to the QCM and to the computer for data recording QCM frequency. In this way the frequency change of the QCM-ZnO structure as a function of time is registered for different NO_2 concentrations.

A schematic diagram of the experimental setup and more details of the methodology for the sensing test are presented in Ref [14]. The sensitivity of the QCM-ZnO system to NO_2 is determined by measuring the frequency – time characteristics (FTCs) at different NO_2 concentrations.

3. Results and discussion

Figure 1 shows Scanning Electron Microscopy (SEM) micrographs of ZnO nanostructured thin film, deposited at constant charge density for 35 min. The structure of this thin film consists of planar hexagonal ZnO nano-whiskers (NW) with thickness of about 100-200 nm and different length (0.5-0.8 μm), stacked with the narrow side to the substrate surface. The SEM micrographs reveal the porous surface morphology of the deposited layer.

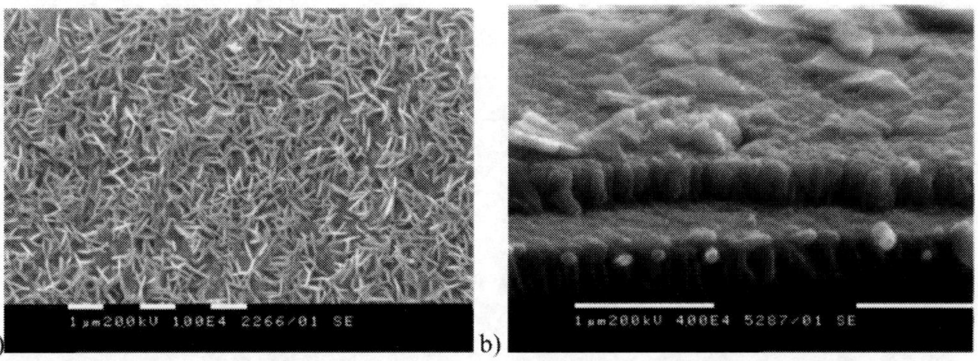

a) b)

Figure 1. SEM micrographs of ZnO nanostructured layer deposited by electrochemical method on quartz resonator: (a) – surface view and (b) cross section view.The markers correspond to 1 μm.

The quartz resonator parameters measured and calculated are: serial frequency F_s = 15.898964 Hz, equivalent dynamic resistance (R_s) = 7.15Ω, and quality factor (Q) = 55 676. After the electrochemical ZnO deposition, as a result of the loading, F_s decreases with 30.29 kHz, while an insignificant worsening is registered for R_s and the Q to 8,2 Ω and 49193, respectively. After treating the QCM-ZnO system at different NO_2 concentrations, the values of its parameters do not change (8.26 Ω and 48943). This indicates that the QCM-ZnO system is stable after the loading processes and it is fully restored after gas treatments in NO_2 concentrations ranging from 100 ppm to 5000 ppm.

The frequency-time characteristics (FTCs) of the QCM-ZnO system measured at different NO_2 concentrations are presented in figure 2. On the basis of the FTCs, the saturation times reached and the corresponding times for system recovering, i.e. desorption times, are determined. The obtained data from the FTCs are presented in figure 3. With the increase of the NO_2 concentration, the total time ($t_{s\ total}$) for saturation during the sorption process increases from 160 sec to 432 sec for gas concentration range 250 ppm - 5000 ppm, approximately three times (2.7). A similar dependence is established also for the desorption process, at which full desorption is reached. The desorption time ($t_{d\ total}$) arises from 100 sec to 244 sec (2.4 times) in the investigated NO_2 concentration range.

Two areas (fast and slow) are observed in the sorption-desorption zone in all measured FTCs. Compared to the full sorption times, $t_{s\ total}$ (160 sec – 432 sec), these belonging to the fast sorption, t_{s1}, are approximately two times shorter – from 85 sec to 188 sec. Comparison of the times of fast sorption (85 sec- 188 sec) to those of fast desorption, t_{d1}, (25 sec -75 sec) shows that the desorption process is twice faster.

From the resulting FTCs curves in all studied gas concentrations it is assumed that the sorption process has an adsorption-diffusion mechanism. The FTCs also show that the sorption process is reversible for all the measured concentrations.

Journal of Physics: Conference Series **559** (2014) 012014 doi:10.1088/1742-6596/559/1/012014

The FTCs make possible not only the characterization of the process of sorption, but also the determination of the maximal quantity of the substance adsorbed at a fixed NO_2 concentration.

The maximal frequency shift of the system during dynamic equilibrium is illustrated in figure 4.

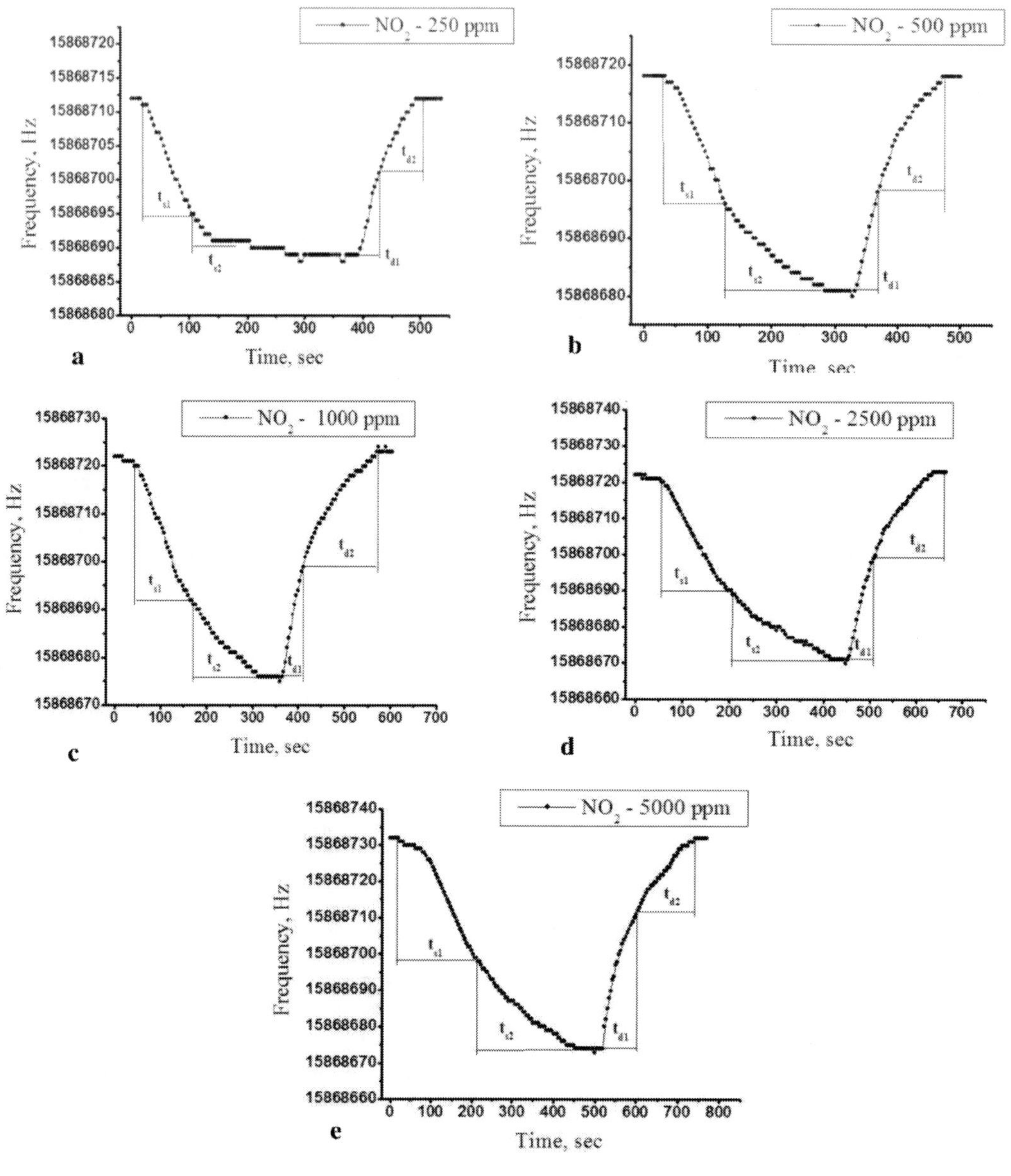

Figure 2 Frequency-time characteristics of QCM with NS ZnO layer at different NO_2 concentrations: a) 250 ppm, b) 500 ppm, c) 1000 ppm, d) 2500 ppm and e) 5000 ppm.

Based on the measured frequency shift, the maximal adsorbed mass is calculated according the Sauerbery equation [15]:

$$\Delta F = -\frac{(2{,}26\times 10^{-6}\times f^2\times \Delta m)}{A}, \tag{1}$$

where ΔF is the measured frequency shift for the time reaching the saturation during the NO_2 exposure (Hz), f - is the frequency of the QCM before exposure to NO_2 (MHz), Δm (g) is the mass of the sorbed gas on the surface of the ZnO films and A = 0.1256 cm^2 - is the surface area of the electrode.

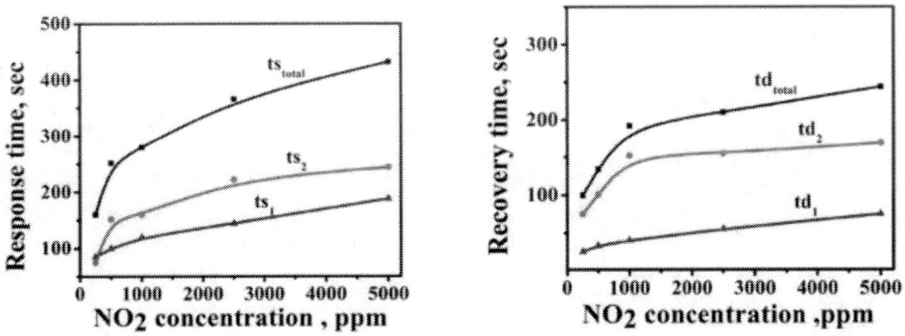

Figure 3. Response and recovery times of QCM-ZnO vs. NO_2 concentration: ts_1 - the time of fast sorption, ts_2 – the time of slow sorption, ts_{tot} = $ts_1 + ts_2$ - the response time; td_1 - the time of fast desorption, td_2 – the time of slow desorption, td_{tot} = $td_1 + td_2$ - the recovery time.

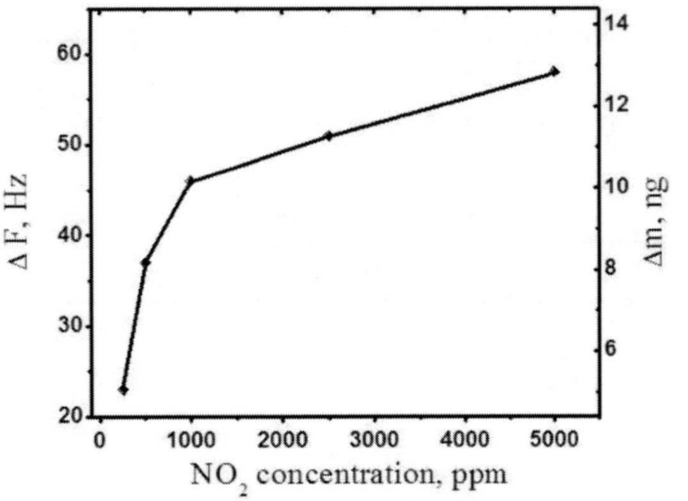

Figure 4. Dependence of QCM-ZnO frequency shift and sorbed mass vs NO_2 concentration.

The obtained dependence is linear for the gas concentration ranging from 250 ppm to 1000 ppm and the change in the frequency shift is from 23 Hz to 58 Hz corresponding to mass change from 5.08 ng to 10.15 ng. The frequency shift values over this concentration decrease significantly, as at 5000 ppm they reach 58 Hz which corresponds to mass of 12,08 ng.

Conclusion

The sensing characteristic to NO_2 of quartz crystal microbalance coated by nanostructured ZnO layer is investigated. The ZnO nanostructured layer is electrochemically grown on Au electrodes of 16 MHz QCM. SEM study reveals that ZnO layer possesses surface consisting of nano-whiskers and has developed surface morphology.

The response and recovery times, maximal frequency shift and sorbed mass at different NO_2 concentrations of QCM with ZnO nanostructured layer are measured. The QCM shows good response to NO_2 at room temperature and can be used for detection of NO_2 concentration higher than 250 ppm. Further more detailed study is required to extend the sensitivity of the system for example by increasing the effective developed surface of ZnO nanostructured layers.

Acknowledgements

The participation of B. Georgieva at the Wokshop on "Transition Metal Oxide Thin Films-Functional Layers in Smart Windows and Water Splitting Devices: Technology and Optoelectronic Properties", 4-6 September, 2014, Varna, Bulgaria, has been supported by the INERA REGPOT Project of the ISSP- BAS and by Project BG051PO001-3.3.07-002 of EU and ESF.

References

[1] Navale S C, Ravi V, Mulla I S , Gosavi S W and Kulkarni S K 2007 *Sensor Actuat B-Chem* **126** 382
[2] Wagh M S, Jain G H, Patil D R, Patil S A and Patil L A 2006 *Sensor Actuat B-Chem* **115** 128
[3] King W H 1964 *Anal. Chem.* **36** 1735
[4] Sun L-X and Okada T 2000 *Anal. Chim. Acta* **421** 83
[5] Nakamura K, Nakamoto T and Moriizumi T 2000 *Mater. Sci. Eng. C* **12** 3
[6] Zhang J, Hu J Q, Zhu F R, Gong H and O'Shea S J 2002 *Sensor Actuat B-Chem* **87** 159
[7] Bunde R L, Jarv E J and Rosentreter J J 1998 *Talanta* **46** 1223
[8] Guibault G G, Hock B and R. Schmid R 1992 *Biosens. Bioelect.* **7** 411
[9] Gomes M, Nogueira P and Oliveira J 2000 *Sensor Actuat B-Chem* **68** 2000 218
[10] Ding B, Kima J, Miyazaki Y and Shiratori S 2004 *Sensor Actuat B-Chem* **101** 373
[11] Wang X H, Ding Y-F, Zhang J, Zhu Z-Q, You S Z, Chen S-Q and Zhu J, 2006 *Sensor Actuat B-Chem* **115** 421
[12] Seiyama T, Kato A, Fujiishi K and Nagatani M 1962 *Anal. Chem.* **34** 1502
[13] Dimova-Malnovska D, Andreev P, Sendova-Vassileva M, Nichev H and Starbova K 2010 *Energy Procedia* **2** 55
[14] Georgieva V, Stefanov P, Spassov L, Raicheva Z, Atanassov M, Tincheva T, Manolov E and Vergov L 2009 *J. Optoelectr. Adv. Mater.* **11** 1363
[15] Sauerbrey G 1959 *Zeitschrift für Physik* **155** 206

Thin Film Microsensing Elements, Technology and Application in Microsystems for Environment Control

G Kolev[1], M Aleksandrova, Y Vucheva and K Denishev
Department of Microelectronic of Faculty of Electronic Engineering and Technologies, Technical University of Sofia, blvd. Kliment Ohridski 8, Sofia

E-mail: georgi_kolev1@abv.bg

Abstract. In this paper microsensing elements with thin piezoelectric films are presented. Three different materials from the basic types – ceramic, metal oxide and polymer, in microsystems for environment parameters control are used.
Zinc oxide (ZnO) and lead zirconium titanate (PZT) by sputtering technology from hot targets are deposited. The optimal parameters of the process and thickness of the layers are presented. Layers of piezoelectrical polymer poly (vinylidene fluoride) (PVDF) were prepared by spray deposition technique.
The microelements (MEMS) with these piezoelectric films are proposed for direct and indirect measuring of parameters as pressure, vibration, gas and liquid sensing, fluid flow, etc.

1. Introduction

The working principle of the piezoelements or structures is always in connection with the presence of a mechanical deformation, that leads to generation of charge carriers or in opposite, by applying a voltage to them, they are mechanically deformed, or a spatial motion of the structure could be observed. That processes are noticed in many MEMS structures and harvesting devices.

The use of the piezomaterials is in connection with the knowledge of their parameters, effects, properties and methods of deposition. The development of technologies and the synthesis of new materials allow successful implementation of structures with piezogenerating or sensitive functions in several electronic devices. New, widely used piezomaterials are: PVDF- organic polymer, PZT-piezoceramic, ZnO- metal oxide etc.

The recent data about poly (vinylidene fluoride) (PVDF), with piezo- and pyroelectrical properties, appears in the specialized literature [1]. The wide application of PVDF, as piezoelectric layers, in different pressure and force sensors and actuators is conditioned from its ability for low cost and simple methods of deposition onto standard substrates.

PZT is piezoelectric from the group of the piezoceramics that could be synthesized by two basic technologies. By the first of them, the corresponding titanates and zirconates are preliminary synthesized and after that mixed in suitable ratio, in order co-synthesis process of the specific component is possible from the solid solvents of PZT. The second technological sequence gives the possibility to produce materials with higher quality. The chemical reactions are given as the following:

$$PbO + TiO_2 = PbTiO_3; \quad PbO + ZrO_2 = PbZrO_3 \tag{1}$$

[1] To whom any correspondence should be addressed.

Journal of Physics: Conference Series **559** (2014) 012015 doi:10.1088/1742-6596/559/1/012015

$$PbTiO_3 + PbZrO_3 = Pb(Zr,Ti)O_3 \quad (2)$$

$$PbO + TiO_2 + ZrO_2 = Pb(Zr,Ti)O_3 \quad (3)$$

ZnO is amongst the most widely studied of all metal oxide systems and has recently become a very popular material. Zinc oxide (ZnO) is a wide band gap (3.37 eV) semiconductor with a large exciton binding energy (60 meV) and one of the most widely used and studied functional oxides[2]. Zinc oxide readily forms into noncentral symmetric wurtzite nanocrystal structures with self-polarized crystal surfaces. Zinc oxide is desirable for many applications, such as piezoelectric transducers, varistors, gas sensors and transparent conducting thin films. ZnO is a key technological material. The lack of a center of symmetry in wurtzite phase, combined with strong electromechanical coupling, results in pronounced piezoelectric and pyroelectric properties and the consequent use of ZnO in mechanical actuators and piezoelectric sensors.

The PZT, ZnO and PVDF layers were investigated by Fourier transform IR spectroscopy (FTIR).

2. Experimental

The PVDF can exist in several crystalline phases. From the five known modifications, the α- and β phases are stable and the most common. The β-phase is the desirable phase due to its piezo-, pyro- and ferroelectric characteristic. For deposition of PVDF layers by spray method, laboratory stand was designed (figure 1), which consists of adjustable flat heater with power 350 W, placed on dielectric ceramic substrate. The set temperature is controlled by electronic regulator. The temperature range which can be set is 30-120°C and the accuracy in this range is 2°C [3].

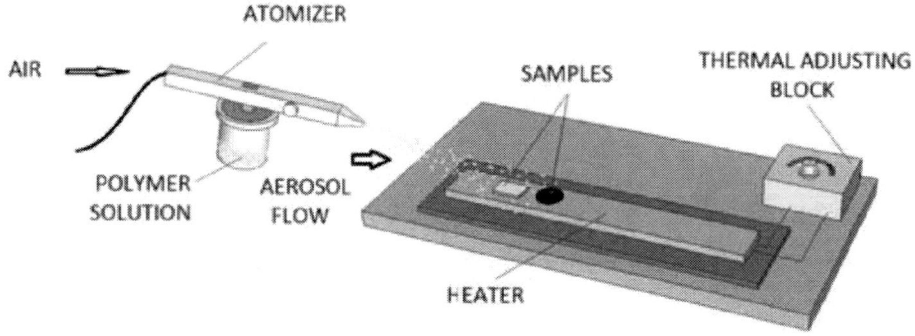

Figure 1. Experimental setup for the spray deposition of PVDF thin films.

The deposition is conducted by atomizer with possibility to regulate the diameter of the nozzle, which enables formation of aerosol flow with different diameters of the droplets. The working air pressure is 3.8 bars. For obtaining of PVDF layers 5 mm diameter grains were used, purchased from Goodfellow. The used solvent for crystallization of PVDF in β-phase is methyl ethyl ketone (MEK).

For the deposition of thin PZT and ZnO layers, the sputtering targets of ($Pb(Zr_{0.52}Ti_{0.48})O_3$) and ZnO with purity 99.99% were purchased from "Semiconductor Wafer, Inc." and "Goodfellow" with 75 mm diameter and 3 mm thick, respectively. PZT target was mounted in vacuum chamber of Leybold sputter coating system A400 VL.

The films were deposited by RF sputtering process. The exact deposition conditions were as follow: base vacuum $2x10^{-5}$ Torr, the gas partial pressure of Ar $2.5x10^{-2}$ Torr, sputtering voltage U_{plsm} = 0.75 kV, I_{plsm} = 150mA, sputtering power P_{plsm} = 112 W. The growth rates are 2.5 nm/min for PZT and 9 nm/min ZnO, respectively [4].

For characterization of the layers, FTIR spectroscopy was applied with IRPresige 21 Shimadzu spectrophotometer in the spectral range of 350 cm^{-1} to 4000 cm^{-1}.

3. Results and discussion

3.1. PVDF layers

For further confirmation of the phases in the PVDF films, FTIR characterization was employed. The FTIR analyzing of layer were in reflectance mode and the measured spectrum is given in figure 2.

According to literature, the absorption peaks at 411, 530, 615, 766, 795, 855, 974, 1149, 1210, 1383, 1402, 1432 and 1455 cm^{-1} can be used for α-phase identification [5, 6].

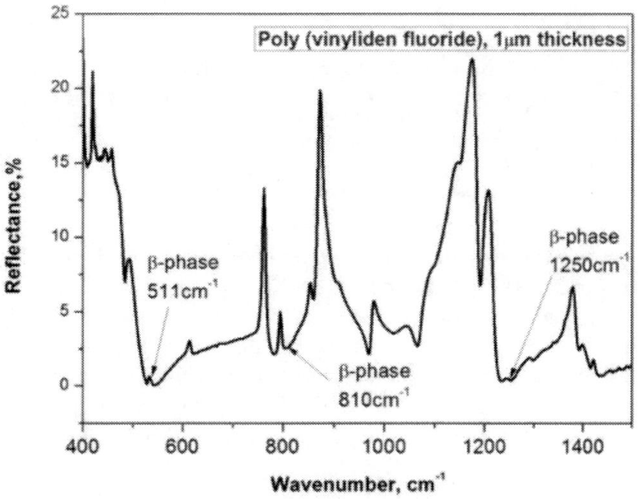

Figure 2. FTIR spectrum of sprayed PVDF film.

The detected peaks at 511, 600, 840 and 1275 cm^{-1} are the fingerprints of β-phase [5, 6]. The identified absorption bands are rich in information on the conformational isomerism of the chain, providing information for the β – phase presence. The characteristic bands related to the β - phase have been identified at 1250, 810 and 511 cm^{-1}.

3.2. ZnO thin films

Infrared spectrum of ZnO films was obtained in reflectance mode in the spectral range of 350 to 1100 cm^{-1} (see figure 3). The absorption band at 390 cm^{-1} is usually reported for ZnO as a typical absorption feature [7].

The absorption peak at 406 cm^{-1} is theoretically confirmed for wurtzite zinc oxide as infrared active mode. The absorption band at 430 cm^{-1} is assigned to the stretching vibrations of Zn-O bond [8, 9], characteristic for ZnO. The IR lines at 560 and 814 cm^{-1} are due to the stretching vibrations of Zn-O bond [10].

FTIR analysis reveals that there is an indication of wurtzite crystal phase in the sputtered zinc oxide films.

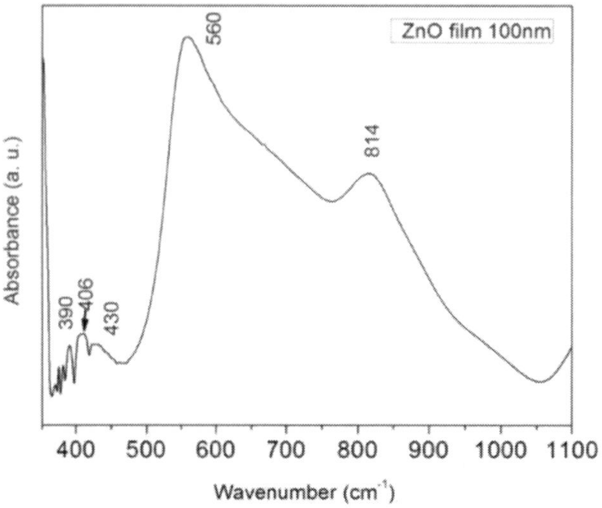

Figure 3. FTIR spectrum of sputtered ZnO film.

3.3. PZT thin films

PZT (and many other piezoelectric materials) have crystal structures belonging to the perovskite family with the general formula $AB0_3$. It is known that the perovskite phase is the only desired crystal structure which has the ideal piezoelectric properties. FTIR spectroscopy as a very sensitive tool can give some information for the chemical bondings and structure of the studied samples.

FTIR spectrum of PZT thin film in reflectance mode is presented in figure 4. The bands appear in the range of from 450 to 350 cm^{-1} are related to B–O vibrations (BO_6 and B–O for ABO_3 structures) [11, 12]. The instinctive feature at 550 cm^{-1} is observed. This band resembles with the reported FTIR observations of PZT ceramics [13].

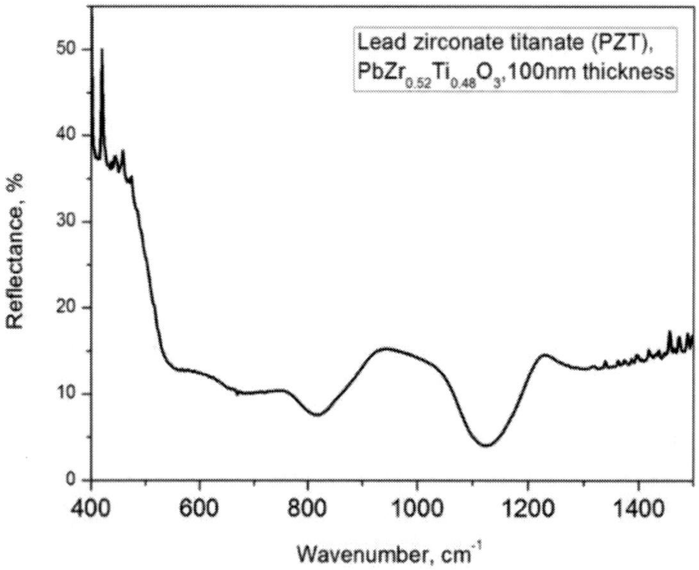

Figure 4. FTIR spectrum of sputtered PZT film.

In this spectral area, a broad band can be seen from 800 to 550 cm^{-1} with a maximum at 780 cm^{-1}. [10] This peak has been associated with the vibration of M–O (M = Zr or Ti) bonds in the systems.

The FTIR results confirm the formation of perovskite phase of lead zirconium titanate. This is a promising result as it manifests that PZT films in perovskite crystal structure can be deposited by RF sputtering process and respectively this is a favorable phase structure for exhibiting good piezoelectric properties.

3.4. Application of the studied films

Many MEMS sensors working with small motion of thin film structures use piezoelectric layers as gas (SAW), pressure, flow sensors, etc. The films in these devices have to possess piezoelectric properties, which are observed by forward and reverse effect. Some typical applications are shown in figure 5.

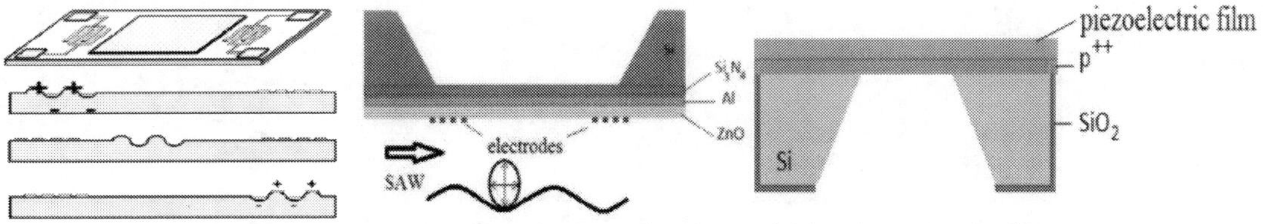

Figure 5. Examples of typical applications of thin piezoelectric films.

4. Conclusion

The layer of PVDF- MEK solution was deposited by spray technology at optimal process parameters. The obtained PVDF films were investigated by FTIR spectroscopy and it was found that the organic solvent MEK contributes for the crystallization of PVDF films in polar β- phase.

Thin films of PZT and ZnO were deposited by RF sputtering process. The FTIR spectrum shows that the perovskite phase is formed in PZT layers by revealing characteristic infrared modes. FTIR spectrum of ZnO film shows the absorption bands that are indicative for wurtzite zinc oxide crystal phase.

The main conclusion that can be drawn from this investigation is that the thin films of PVDF, PZT and ZnO can be successful candidates as piezoelectric layers in MEMS devises, sensors and actuators for environment control.

Acknowledgment

This paper was presented at INERA Workshop "Transition Metal Oxide Thin Films-functional Layers in "Smart Windows" and Water Splitting Devices", September 4-6th, 2014, Varna, Bulgaria. The Workshop is part of the Program of INERA REGPOT Project of Institute of Solid State Physics, Bulgarian Academy of Sciences.

The authors are thankful to Assoc. Prof. Dr Tatyana Ivanova (Central Laboratory of Solar Energy and New Energy Sources at the Bulgarian Academy of Science) for the measurements of FTIR spectra.

References

[1] Vijayakumar P, Khakhar D V and Misra A 2010 *J. Appl. Polymer Sci* **117** 3491–3497
[2] Djelloul A, Aida M-S and Bougdira J 2010 *J. Lumin.* **130** 2113–2117
[3] Kolev G, Aleksandrova M and Denishev K 2013 *IJRIS* **5** 177-182
[4] Kolev G, Denishev K, Aleksandrova M and Dutcolova Y 2013 *ICEST* **2** 691-694
[5] Lanceros-Mendez S, Mano J F, Costa A M and Schmidt V H 2011 *Sci. Phys. B* **40** 517–527
[6] Satapathy S, Gupta P K, Pawar S and Varma K B R 2011 *Bull. Mater. Sci.* **34** 727–733
[7] Li C P, Jang B H, Wang X C, Li M J, Su L and Li X W 2011 *Appl. Surf. Sci.* **257** 5998-6003
[8] Muthukumarana S and Gopalakrishnan R. 2012 *Optic. Mater.* **34** 1946–1953

Journal of Physics: Conference Series **559** (2014) 012015 doi:10.1088/1742-6596/559/1/012015

[9] Verma A, Kar M and Agnihotry S A 2007 *Solar Energy Mater. Solar Cells* **91** 1305
[10] Pakizeh E, Hosseini S M, Kompany A and Ghasemifard M 2010 *Inter. J. Nanosci.* **3** 193-199
[11] Pontes F M, Leit E R, Nuns M S J, Pontes D S L, Longo E, Magnani R, Pizani P S and Varela J A 2004 *J. European Ceramic Soc.* **24** 2969-2976
[12] Zak A Kh and Majid W H Abd 2010 *Ceram. Inter.* **36** 1905–1910
[13] Kumari N, Singh A and Mandal J 2013 *Inter. Refer. J. Engin. Sci. (IRJES)* **2** 48-50

Transparent Conductive Oxides in Thin Film Photovoltaics

Frank U Hamelmann

University of Applied Sciences Bielefeld, Technical Faculty, Artilleriestraße 9a, 32427 Minden, Germany

frank.hamelmann@fh-bielefeld.de

Abstract. This paper show results from the development of transparent conductive oxides (TCO's) on large areas for the use as front electrode in thin film silicon solar modules. It is focused on two types of zinc oxide, which are cheap to produce and scalable to a substrate size up to 6 m^2. Low pressure CVD with temperatures below 200°C can be used for the deposition of boron doped ZnO with a native surface texture for good light scattering, while sputtered aluminum doped ZnO needs a post deposition treatment in an acid bath for a rough surface. The paper presents optical and electrical characterization of large area samples, and also results about long term stability of the ZnO samples with respect to the so called TCO corrosion.

1. Introduction

For efficient thin film solar cells transparent conductive oxides (TCO's) are needed as a front side electrode, since a metallic grid as used in crystalline silicon solar cells cannot be used. Various metal oxides with different doping materials can be used as a TCO. In general, a tradeoff between transparency and conductivity of the TCO must be made. For highest conductivity at lowest optical absorption sputtered indium tin oxide (ITO) is an established thin film material on glass, which can be bought on glass substrates in many sizes. However, this material is expensive and should be replaced for low cost applications such as thin film solar modules. Another established TCO is fluorine doped tin oxide (FTO), which is produced by glass companies in an APCVD process in the floatline. It is available on soda-lime glass and on low-iron glass with a substrate size up to 6 m^2 (Nippon Sheet Glass). This TCO is important for thin film silicon solar modules, since the FTO shows a native substrate roughness with pyramidal shapes, which is important for the light management in the amorphous silicon. In difference to other thin film solar cell technologies, thin film silicon requires besides high optical transparency and high electrical conductivity a rough surface that is able to scatter light under large angles, without disturbing the growth of the silicon with sharp edges or pinholes.

This paper will show results for another class of transparent conductive oxides, based on zinc oxide. Large area deposition was introduced in the 1980s for Ag-based low emissive coatings (low-E) in energy efficient glazing [1]. Reactive DC sputtering is still today used for these dielectric coatings in architectural glazing [2, 3]. These developments were the basics for the high rate magnetron sputtering of aluminum doped zinc oxide (AZO) [4, 5]. The process of high rate sputtering following by a wet chemical etching to achieve a tailored surface roughness suitable for thin film silicon solar cells was (amongst others) developed at the Institute for Photovoltaics of the Forschungszentrum Jülich, the latest results with a two-step etching process with HCl and HF are described in [6]. The etching creates craters on the surface with the ability of excellent light scattering, as it is needed for high efficiency thin film silicon solar cells. Applied Materials commercialized the sputter/etch process for their "SunFab" turn key thin film silicon production line [7, 8], which was able to produce solar modules with a size of up to 6 m^2.

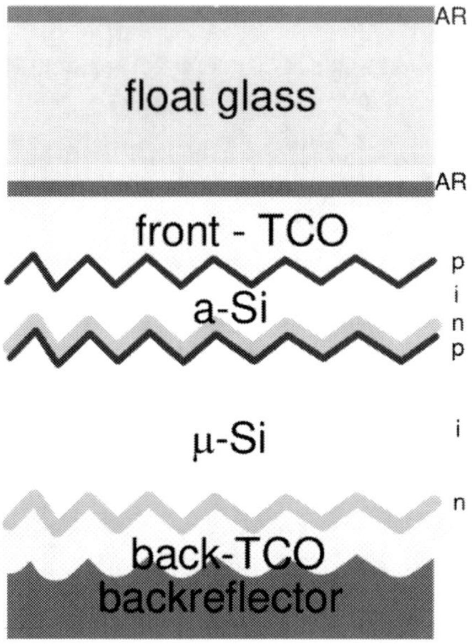

Figure 1. Layers of a thin film silicon solar cell.

The second large area deposition strategy for doped ZnO is the low pressure CVD process (LPCVD). The first developments were patented by Siemens Solar in the 1980's for their CIS thin film solar cells [9, 10]. The University of Neuchatel spent a lot of work to create a low pressure process with boron doping, based on the reaction of diethyl zinc (DEZ) and water vapor at moderate substrate temperatures [11]. This technology was licensed by Oerlikon Solar for their turn key thin film silicon production line "ThinFab" for solar modules of 1.4 m^2 size [12]. The process delivers a transparent conductive oxide with a pyramidal shaped surface texture that is excellent for light scattering in thin film silicon solar cells. The same process can be used for the creation of the cells back contact. Combined with a white paint as back reflector, this is a highly efficient alternative to the usually used metallic back contact of thin film silicon solar cells. Figure 1 shows the layers of a typical thin film silicon solar cell, with a tandem structure of amorphous (a-Si) and microcrystalline (μ-Si) silicon.

2. Methods

The presented results were obtained in the R&D facilities of the Malibu GmbH & Co KG in Bielefeld, Germany. The boron doped ZnO was deposited on soda lime glass in the TCO-1200 LPCVD system from Oerlikon Solar, the R&D version of their commercial deposition system. The System contains three chambers: a loadlock with pre-heating of the substrate, the deposition chamber and the unloadlock for controlled cooling down of the substrates. DEZ and H_2O were vaporized and through mass flow controllers introduced to the deposition chamber. B_2H_6 diluted in H_2 was used for doping. To achieve a uniform temperature distribution on the substrate surface in a temperature range between 150 and 200°C, the substrate plate was divided into three zones with separate heating control. To avoid glass damage the pre-heating and post deposition cooling had to be done careful. For reproducible results, before deposition runs the system was conditioned with 5 runs on dummy substrates. The thickness of the films was controlled by deposition time.

The sputtered aluminum doped zinc oxide was prepared in a sputtering system of Leybold Optics, consisting of a load lock chamber, a deposition chamber for metal films (silver and aluminum), and a deposition chamber for oxide films (zinc oxide and SiO_2). The zinc oxide was DC sputtered with rotatable ZnO:Al sputter targets. The substrates (soda lime glass, 1.4 m^2) were heated up to 400°C in the loadlock, and transported to the process chamber where the substrate temperature was kept at 400°C. The thickness of the films was controlled by the speed of the substrate over the sputter target, and the number of sputter cycles. After deposition, the ZnO coated substrates where processed in a modified glass washer from RENA with a separate chamber containing highly diluted hydrochloric acid at 25°C. The speed of the substrate in the chamber and the pH value of the HCl solution give the etching intensity.

Standard characterization of the TCO-films was performed with a self constructed x-y table with optical film thickness measurement (Micropack Nanocalc spectroscopic reflectometer) and 4-point probe (Jandel RM3) for spatially resolved sheet resistance. For further analysis, the substrates where cut to smaller pieces and analyzed with UV-vis-NIR spectrometry for transmission, reflection and absorption. The spectrometer was equipped with an Ullrich sphere for wavelength resolved measurement of the haze value (amount of diffuse transmitted light), which is important for the light scattering. For some detailed material analysis, Raman spectroscopy was used, also FTIR spectroscopy, atomic force microscopy, scanning electron microscopy and other methods.

For the R&D of TCO in thin film solar cell production the main challenge is achieving higher efficiency at lower costs. For this the following topics can be highlighted:

- Improving the material uniformity on large areas
- Reducing the total processing time
- Reducing he raw material consumption
- Improving the basic understanding of the material and the processes
- Assuring the long term stability under operating conditions

3. Results

For the use in 1.4 m^2 solar modules the homogeneity of the TCO is the most important issue. To get an idea about the distribution of film thickness and sheet resistance, both values are measured on a x-y table with a distance 10 cm for each point.

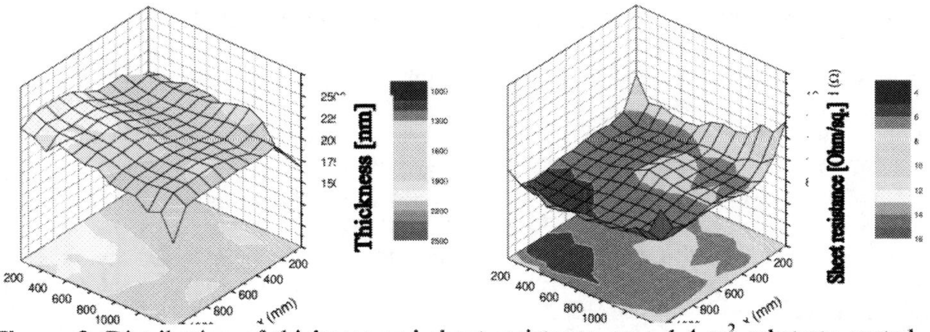

Uniformity
33.9%
ρ_{max} (Ω)
11.4
ρ_{min} (Ω)
5.6
ρ (Ω)
6.9
$\Delta\rho$ (Ω)
5.8
ρ_{median} (Ω)
6.7
σ (Ω)
1.0

Figure 2. Distribution of thickness and sheet resistance on a 1.4 m^2 substrate coated with ZnO:B. Statistical data of the sheet resistance distribution are given on the right side.

Journal of Physics: Conference Series **559** (2014) 012016 doi:10.1088/1742-6596/559/1/012016

An automated detailed statistical analysis of the results makes it easy to decide if the distribution is good enough for the purpose. Figure 2 shows an example of a LPCVD produced ZnO:B layer. It can be seen, that the coating on the corners is the too thin, resulting in a too high sheet resistance. A carefully adjusted temperature distribution of the hot plate with its three heating zones is essential for a good homogeneity.

In order to do advanced measurements, the substrates have to be cut into smaller pieces. Figure 3 shows on the left graph the results of a series of LPCVD samples produced with the same parameters at the same growth rate, but at different deposition times. To exclude effects from inhomogeneous thicknesses on the substrate, each sample was taken from the same position on the glass. An almost linear behavior can be seen. Also the development of the sheet resistance is proportional to 1/d as it should be. It must be noted that the thickness estimation is based on reflectivity measurements it needs the proper optical constants to deliver good results. So the resistance measurement is usually more reliable to determine the uniformity of the TCO.

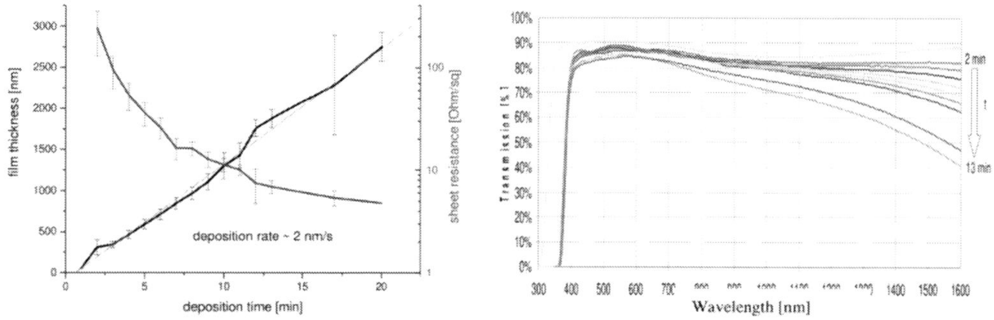

Figure 3. Film thickness, sheet resistance (left) and transmission (right) at various deposition times

On the right side the corresponding measurements in the UV-vis spectrometer show the increasing absorbance of the films with increasing film thickness. Especially in the IR the absorbance increases strongly. It must be noted, that for amorphous silicon the useable wavelength is between 400 and 700nm, while (micro-) crystalline silicon can convert light up to 1100 nm into electrical energy. Such measurements can also be used to calculate the important wavelength dependent "haze", the amount of diffuse transmitted light to the total transmitted light (not shown). Many other detailed observations can be made on such small samples to explain the microscopic structure of the material and its electrical, optical and morphological properties and the influence of the deposition parameters on the material. However, the ultimate test for the quality of a TCO sample is the fabrication of cells of complete modules.

Another example of an important parameter for the large area homogeneity of LPCVD produced ZnO:B is shown on figure 4. In this graphs the distribution of the sheet resistance on the full 1.4 m^2 area of the glass substrate is illustrated for two different samples. All parameters were the same during deposition, except the total pressure in the chamber, adjusted by a butterfly valve. While the uniformity is good on the left graph for the sample produced at 1.0 mbar, a lower pressure of 0.5 mbar results in a higher average resistance and a higher standard deviation.

The same experiments as above were performed for sputtered and etched aluminum doped ZnO samples. The distribution of the sheet resistance looks different to the LPCVD samples, and it is even more difficult to achieve good uniformity since two completely different processes are involved. Figure 5 shows the sheet resistance on a 1.4 m^2 substrate with an AZO thickness of 1000 nm, measured after deposition.

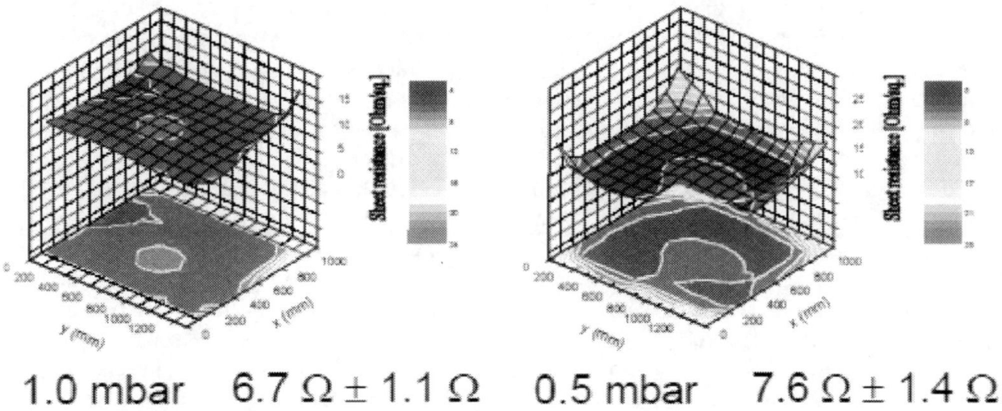

Figure 4. Distribution of the sheet resistance at various deposition pressure.

It can be seen that the edges in the direction of the movement in the sputter system (right and left edge on the graph) show a lower resistance than the center. The following etch step makes things even worse, since the etching rate depends on the material a lower uniformity can be expected after etching. The process development for this two-step process takes a lot of time and many runs on full size systems.

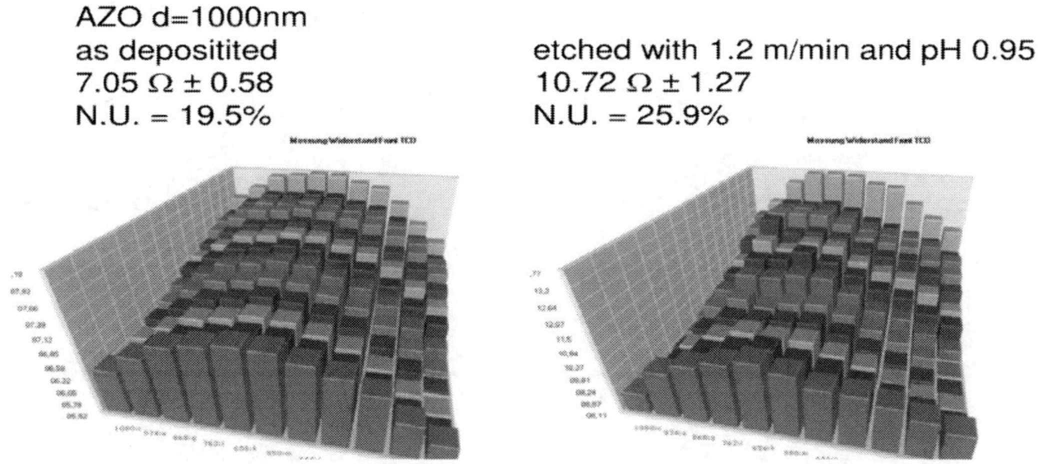

Figure 5. Distribution of sputtered AZO sheet resistance before and after etching (N.U. = non-uniformity)

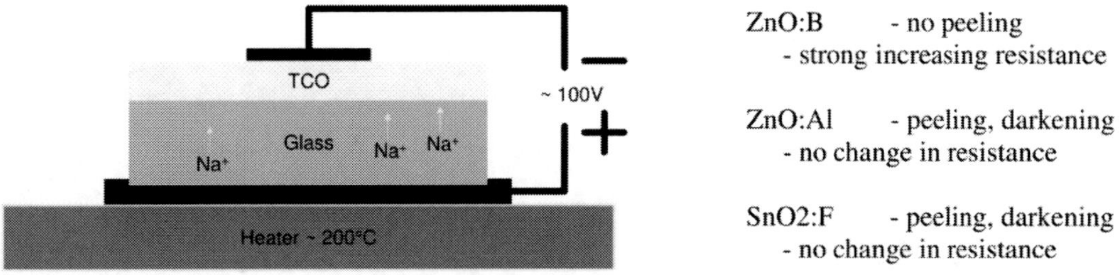

Figure 6. Accelerated test for TCO corrosion, results.

A very important issue for production quality is the long term stability of modules. Besides the efficiency of the encapsulation one important aspect for thin film modules is about the TCO-glass interface. During operation of the modules a potential between the TCO and the frame of the module of several hundred volts may occur. At high operating temperatures of up to 70 °C the migration of sodium ions from the glass into the TCO can happen, resulting in a destroyed TCO layer after a partly couple of years. This co called "TCO corrosion" was first investigated for SnO2:F coatings in a-Si modules [13], but it is also important for ZnO coatings. It can be suppressed by avoiding frames and/or using inverters with galvanic separation, or using sodium free glass. However, all this is a disadvantage for the system, so other solutions like diffusion barriers may be an option. For accelerated corrosion tests, the TCO sample is contacted at the glass side with a conductive glue. Between TCO and glass a voltage of about 100V is applied, while the whole sample is heated up to 200°C. A small current of some µA can be observed. After 15 minutes the test is finished, and the damage to the TCO can be seen.

Here some major difference between ZnO:Al, SnO_2:F and ZnO:B can Here some major difference between ZnO:Al, SnO_2:F and ZnO:B can be seen. While the first two materials show a massive cracking on large parts of the film, but no change in resistance at the undamaged parts, the ZnO:B seems to be undamaged, but the whole film has a highly decreased resistance. In all cases it could be shown, that a silicon dioxide layer of some 10 nm is suitable to reduce the corrosion effects by several orders.

To explain the different behavior in the corrosion test, grazing incidence x-ray diffraction (GIXRD) was used to determine the stress in the films (see figure 7). ZnO:B shows no stress in the bulk film, while SnO_2 exhibits tensile stress of 221 MPa, and ZnO:Al high compressive stress of 646 MPA on the glass substrate.

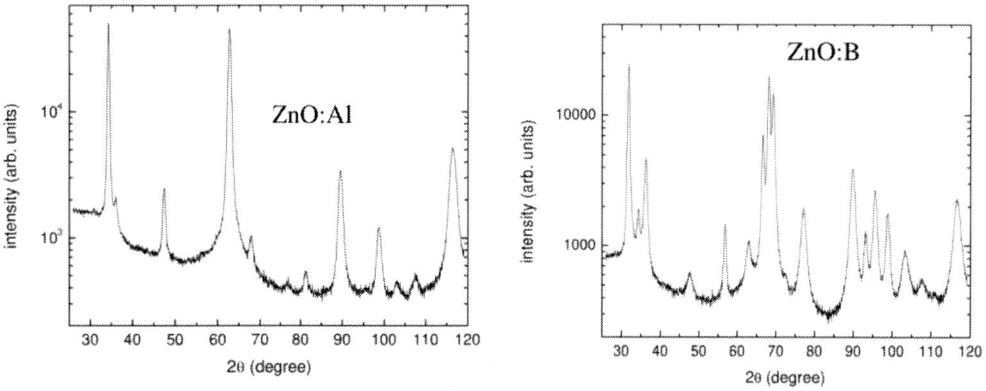

Figure 7. GIXRD measurement on ZnO:Al film (left) and ZnO:B (right) under 1.5°

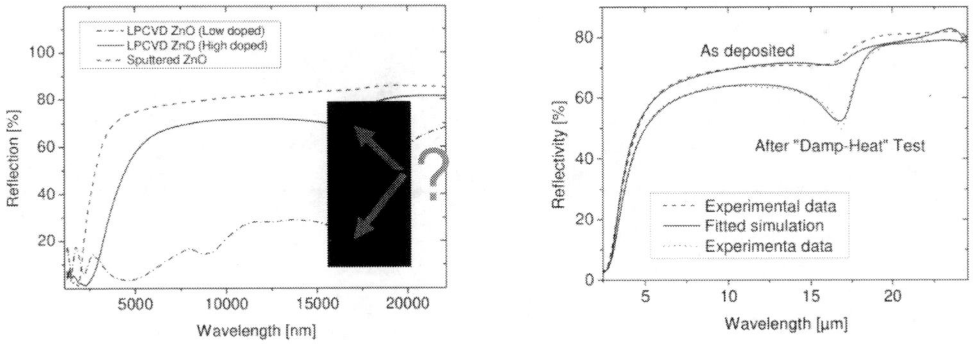

Figure 8. IR reflection spectra of different ZnO (left), measured and calculated IR spectra for damp-heat test

Further experiments explained the structure of the ZnO:B and the increasing of the resistance. It is well known, that ZnO:B is sensitive to humidity [11, 12]. However, the work at Malibu showed that the LPCVD produced ZnO:B shows a connected network of voids, allowing the diffusion of ions through the bulk material [14] much easier than in more compact material like sputtered ZnO:Al. In addition, it could be shown in [15] that the results of IR reflectivity obtained structures at high wavelength (between 15 and 20 µm) that could not be explained by the standard Drude model (figure 8, left graph).

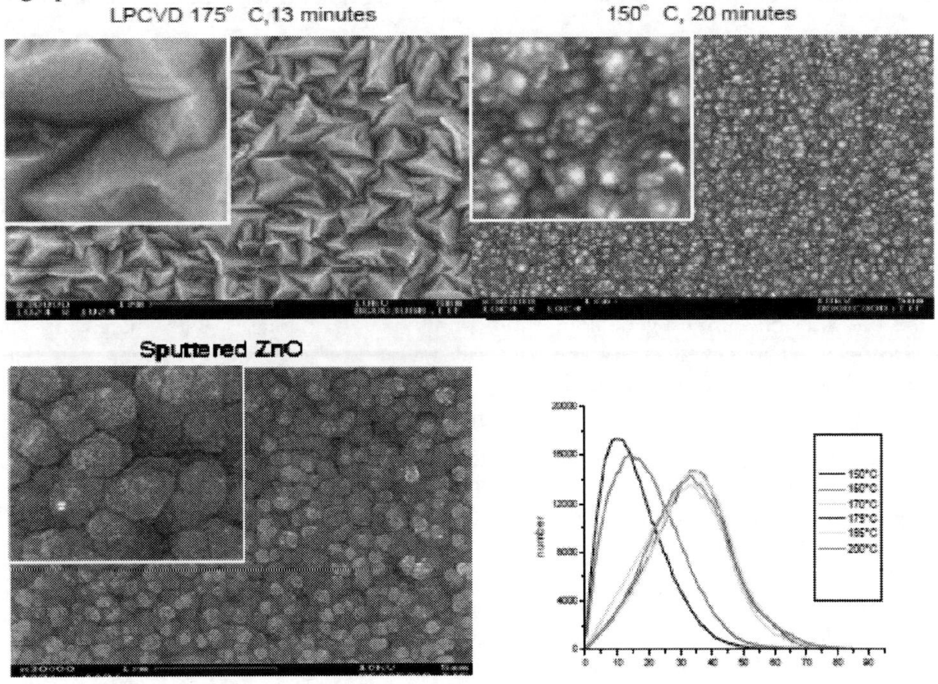

Figure 9. SEM images of ZnO surfaces, results of AFM imaging (lower right).

The Maxwell-Garnett model however could describe the measurements very well with a depletion layer around the ZnO grains. It can be seen on the right graph of figure 8, that the IR reflection spectra after a damp-heat test (at 80°C and 80% humidity) can be simulated with an increased depletion layer.

Figure 9 shows SEM images of ZnO:B produced at 175°C (upper left), 150°C (upper right) and sputtered ZnO:Al (lower left). It is obvious, that the structure of the 175°C ZnO:B is completely different to the other samples, it shows large pyramidal shapes as needed for light scattering. The analysis of AFM images on LPCVD samples produced at different temperatures shows a transition between 160°C and 170°C in the angle of the surface elements (figure 9, lower right). Additional analysis with XRD showed, that at temperatures of 160°C and lower, the ZnO:B with a uniform orientation in the (002)-plane (c-axis perpendicular to the surface), just like the ZnO:Al does. At temperatures between 170°C and 190°C the crystal orientation is uniform in the (110)-plane (c-axis parallel to the surface), resulting in the pyramidal shapes [11].

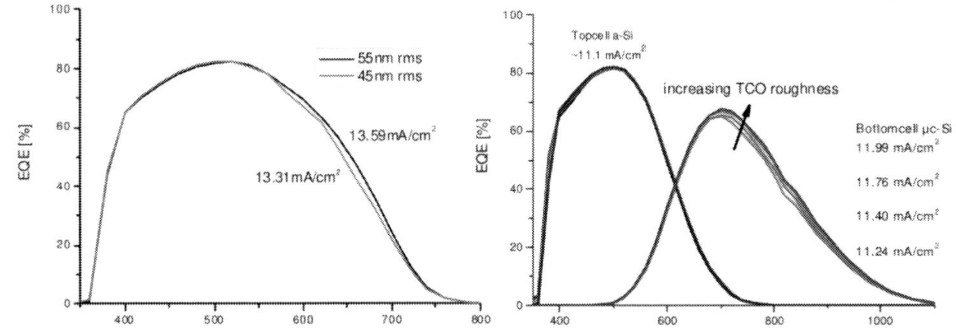

Figure 10. External quantum efficiency of a-si (left) and tandem cells (right) on different ZnO substrates.

As mentioned before, the ultimate test for a TCO is the measurement of the cell or module efficiency in a solar simulator. For the optical performance of a TCO in the cell, the external quantum efficiency is an important tool. Figure 10 shows on the left side the EQE of an a-Si cell on two different ZnO:B substrates. It can be seen, that a higher surface roughness results in a higher current generated between 550 and 700 nm. A similar experiment is shown on the right side for a tandem cell structure (a-Si/µc-Si), the current generated in the bottom cell increases with increasing ZnO:B roughness. However, the generated voltage and fill factor are also important for the resulting efficiency of a cell, and limit the surface roughness.

4. Conclusions

This paper showed some important examples of the development of ZnO as a transparent conductive layer in industrial thin film solar module production. Large area uniformity is as important as the basic understanding of the material and the processes in order to improve the module performance with respect to the production costs.

Acknowledgments

The author thanks the former Malibu R&D team, especially Henning Kurz and Dr. Pavel Prunici.

References

[1] Gläser H J 1980 *Glass Technol.* **21** 254
[2] Arbab M 2001 *Thin Solid Films* **381** 15

[3] Tsuda Y, Omoto H, Tanaka K and Ohsaki H 2006 *Thin Solid Films* **502** 223
[4] Ellmer K, Kudella F, Mientus R, Schieck R and Fiechter S 1994 *Thin Solid Films* **247** 15
[5] Müller J, Schöpe G, Kluth O, Rech B, Sittinger V, Szyszka B, Geyer R, Lechner P, Schade H, Ruske M, Dittmar G and Bochem H-P 2003 *Thin Solid Films* **442** 158
[6] Jost G, Merdzhanova T, Owen J I, Zimmermann T and Hüpkes J 2014 *Progress in Photovoltaics: Research and Applications,* accepted
[7] Klein S, Rohde M, Buschbaum S and Severin D 2012 *Solar Ener. Mater. Solar Cells* **98** 363
[8] Klein S, Wieder S, Buschbaum S, Rohde M, Schwanitz K, Stolley T, Stömmer C, Severin D, Straub A, Schmidt U I and Ahmed K 2011 *Phys. Status Solidi C* **8** 2978
[9] Wieting R D and Potter R R 1986 **U.S. Patent** 4,612,411
[10] Vijayakumar P S, Blaker K A, Wieting R D, Wong B, Halani A and Park C 1988 **U.S. Patent** 4,751,149
[11] Fay S, Shaw A 2008 ed Ellmer K, Klein A, Rech B *Transparent Conductive Zinc Oxide*, Springer Series in Material Science **104**, 235
[12] Kroll U, Meier J, Benagli S, Borello D, Steinhauser J, Vallat E, Castens L, Orhan J, Djeridane Y, Multone X, Fesquet L, Fecioru-Morariu M, Mereu B, Kalas J, Hoetzel J, Losio P, Kupich M and Kluth O, Eisenhammer T 2011 *Proc. 37th IEEE Photovoltaic Specialists Conference*, 213
[13] Osterwald CR, Mcmahon T J and del Cueto J A 2003 *Solar Ener. Mater. Solar Cells* **79** 21
[14] Beyer W, Breuer U, Hamelmann F, Hüpkes J, Stärk A, Stiebig H, Zastrow U 2009 *MRS Proc.* **1165**
[15] Prunici P, Hamelmann F, Beyer W, Kurz H and Stiebig H 2013 *J. Appl. Phys.* **113** 123104

Journal of Physics: Conference Series **559** (2014) 012017 doi:10.1088/1742-6596/559/1/012017

Parameter variation of the one-diode model of a-Si and a-Si/μc-Si solar cells for modeling light-induced degradation

J A Weicht, F U Hamelmann and G Behrens

University of Applied Sciences Bielefeld, Technical Faculty, Artilleriestraße 9a, 32427 Minden, North Rhine-Westphalia, Germany

E-mail : JWeicht@FH-Bielefeld.de

Abstract. For analyzing the long-term behavior of thin film a-Si/μc-Si photovoltaic modules, it is important to observe the light-induced degradation (LID) in dependence of the temperature for the parameters of the one-diode model for solar cells. According to the IEC 61646 standard, the impact of LID on module parameters of these thin film cells is determined at a constant temperature of 50°C with an irradiation of 1000 W/m² at open circuit conditions. Previous papers examined the LID of thin film a-Si cells with different temperatures and some others are about a-Si/μc-Si. In these previous papers not all parameters of the one-diode model are examined. We observed the serial resistance (Rs), parallel resistance (Rp), short circuit current (Isc), open circuit voltage (Uoc), the maximum power point (MPP: Umpp, Impp and Pmpp) and the diode factor (n). Since the main reason for the LID of silicon-based thin films is the Staebler Wronski effect in the a-Si part of the cell, the temperature dependence of the healing of defects for all parameters of the one-diode model is also taken into account. We are also measuring modules with different kind of transparent conductive oxides: In a-Si thin film solar cells fluorine-doped tin oxide (FTO) is used and for thin film solar cells of a-Si/μc-Si boron-doped zinc oxide is used. In our work we describe an approach for transferring the parameters of a one-diode model for tandem thin film solar cells into the one-diode model for each part of the solar cell. The measurement of degradation and regeneration at higher temperatures is the necessary base for optimization of the different silicon-based thin films in warm hot climate.

1. Introduction

Silicon-based thin film solar cells have a transparent conductive oxide of doped zinc or tin oxide as front contact. Amorphous silicon-based (a-Si) solar cells consist of a thin film p-i-n structure of amorphous silicon (a-Si). Micromorphous thin film solar cells consist of a tandem structure of an amorphous and a microcrystalline silicon thin film (a-Si/μc-Si): each has a p-i-n structure. In the presentation of our work we want to show the electrical behavior of a compound of these different kinds of thin film cells under different climate conditions.

The Staebler - Wronski effect describes the degradation of amorphous silicon (a-Si) solar thin film cells. The a-Si solar cell degradation depends strongly on operating temperature. [1]. Because of their amorphous layer, light-induced degradation is also observed in amorphous/microcrystalline tandem (a-Si/μc-Si) solar cells. [2]

All current-voltage-(I-V) curves can be well described by the extended single-diode model, as seen in Fig. 1. For a tandem solar module, like a-Si/µc-Si, one single-diode model must be selected for each of the two solar cells.

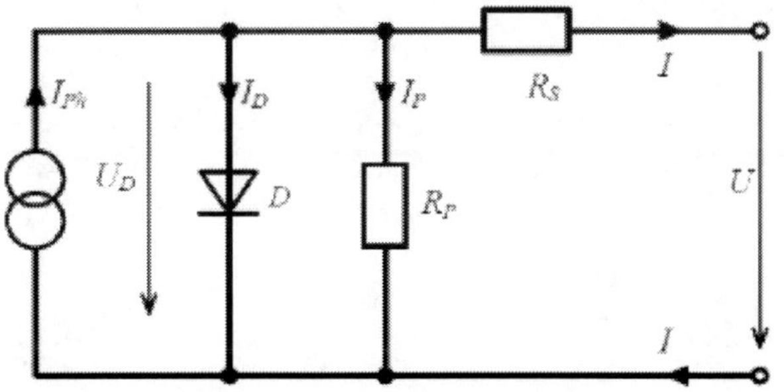

Figure 1. One-diode-model of a solar module.[3]

In our work we are observing the dependence of the serial resistance (Rs), the parallel resistance (Rp), the short circuit current (Isc), the open circuit voltage (Voc) and the diode factor (n) on light-induced degradation and the procedure of annealing of thin-film silicon-based solar cells connected with zinc oxide as described by the Staebler-Wronski effect.

In this equivalent circuit the serial resistance describes, among other ohmic loses, the internal losses of electrical energy through the recombination of electron-hole pairs at the dangling bonds, which are caused by the Staebler-Wronski effect [4]. This recombination in dangling bonds increases during the light-induced degradation [5]. The creation of these dangling bonds is reversible [6]: As described by the Staebler-Wronski effect (SWE), the amount of dangling bonds reduces at higher temperatures. Consequently the probability of recombination of electron-hole pair reduces with higher temperatures. The efficiency at standard test conditions (STC) and the fill factor decrease with increasing amount of dangling bonds and vice versa.

2. Materials and Methods

Solar cells of a-Si and a-Si/µc-Si are degraded under controlled conditions at different constant temperatures and constant irradiation of 1000W/m². At fixed times the efficiency of the modules is measured at 25°C, 1000W/m² and a spectrum of AM 1.5, so called standard test conditions (STC). Consequently comparable values are received. The key parameters n, Voc, Isc, Rs, and Rp are extracted of the measured I-V-curves.

The samples are placed under such a light-soaking bench (LSB) for several days. The sun simulator has an irradiation of 1000W/m². The modules are degraded usually at 50°C. For our measurements the LSB is set to a constant temperature of 70°C, 50°C and 30°C. The mini modules are retained in regular time intervals from the LSB, cooled down to 25°C and measured again at STC. Afterwards they are put back to the LSB for further light soaking.

The diode factor was calculated by the measured I-V-curve for V>Vmpp with elementary charge (e), Boltzmann constant (k) and the temperature of the solar module (T):

$$Nn = \frac{e}{kT}\frac{dV}{d|I_{dark}|} \tag{1}$$

with

$$I_{dark} = I_{sc} - I - \frac{V - IR_s}{R_p} \tag{2}$$

N is the number of measured solar cell: for tandem solar cell (two solar cells) is N=2.

3. Results

3.1. Degradation of a-Si solar cells at different temperatures

As it was already reported by Staebler and Wronski [1], the efficiency and the fill factor of amorphous thin film solar cells decrease in the first 100 hours in dependence of the constant operating temperature. Two components of the fill factor are Voc and Isc. During the light-induced degradation Voc and Isc are decreasing: The lower the degradation temperature the lower the stabilized Voc and Isc. This decreasing effect is shown in figure 2 and 3.

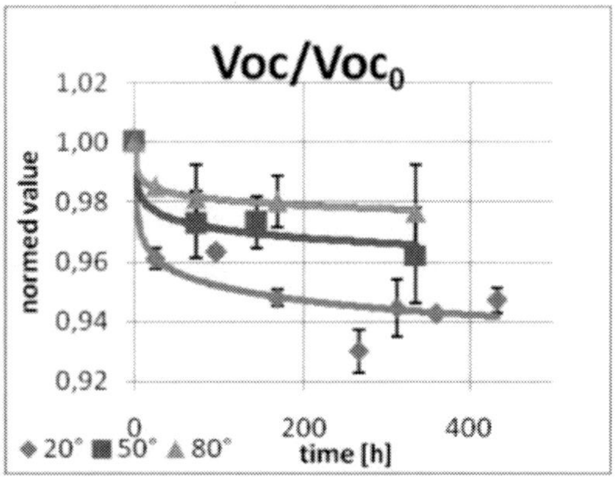

Figure 2. Open circuit voltage Voc of a-Si during light-induced degradation normed at its initial value (Voc_0)

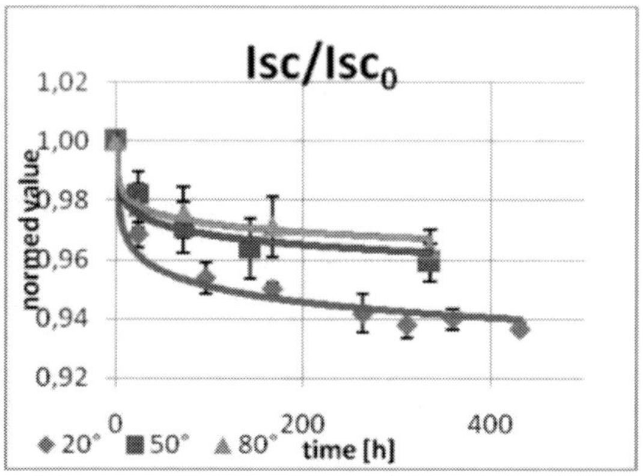

Figure 3. Short circuit current Isc of a-Si during light-induced degradation normed at its initial value (Isc_0).

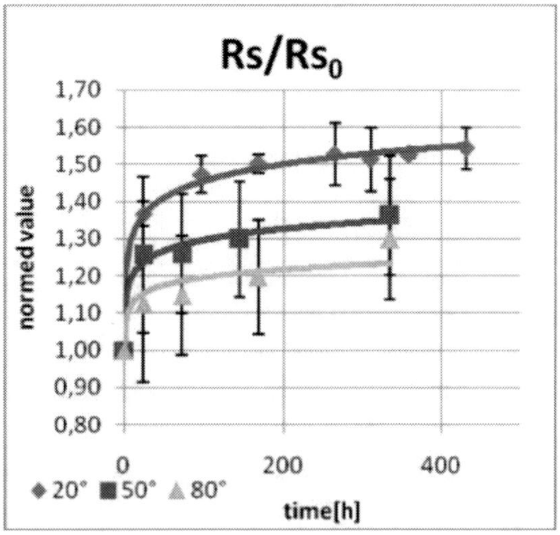

Figure 4. Serial resistance Rsof a-Si during light-induced degradation normed at its initial value (Rs_0)

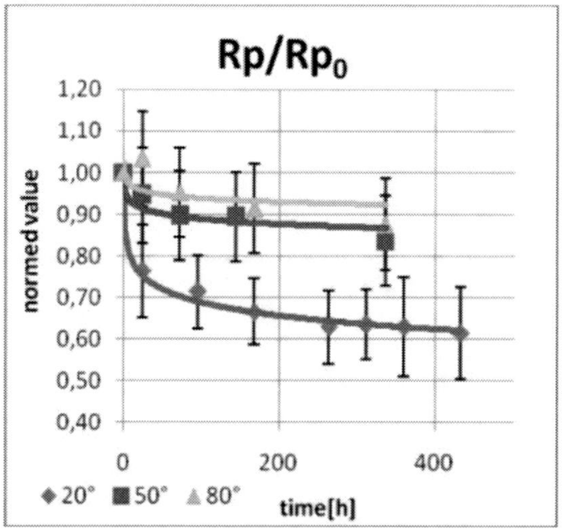

Figure 5. Parallel resistance Rp of a-Si during light-induced degradation normed at its initial value (Rp_0).

The different resistances Rs (figure 4) and Rp (figure 5) have got an opposite trend: Rs increases, while Rp decreases. The impact of the variation of these parameters is a function of the decreasing temperature: The higher the temperature the higher is the impact of the light-induced degradation.

3.2. Degradation of a-Si/μc-Si solar cells at different temperatures

The degradation of a-Si/μc-Si is caused by the degradation in the a-Si solar cell. The degradation is proportional to the thickness of the intrinsic layer of the a-Si layer [7]. To eliminate the different thickness in the intrinsic layers caused by production fluctuations, the solar cells were first degraded at 50°C and afterwards we degraded them at a lower temperature. All of the values were normed of their initial value before the degradation at 50°C.

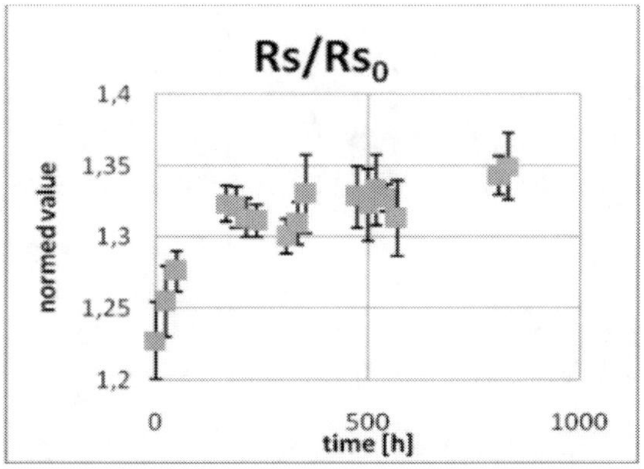

Figure 6. Rs of a-Si/μc-Si after stabilization at a degradation temperature at 50°C and during light-induced degradation at 30°C normed at its initial value (Rs_0)

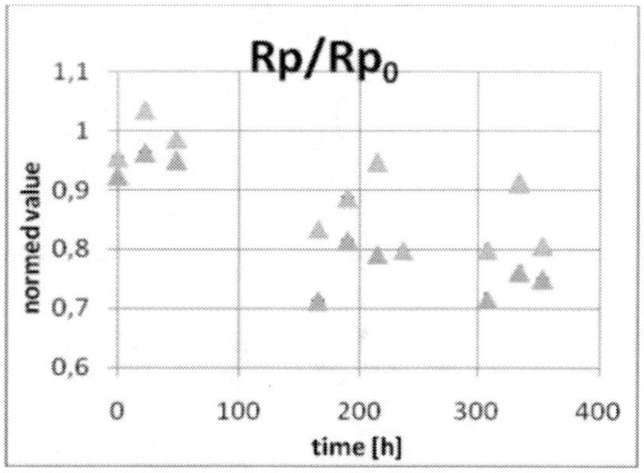

Figure 7. Rp of a-Si/μc-Si after stabilization at a degradation temperature at 50°C and during light-induced degradation at 30°C normed at its initial value (Rp_0).

The serial resistance increases during the degradation at a temperature of 50°C at a factor of 1.22 of the initial value. Afterwards the samples were degraded at 30°C: the serial resistance increases further (figure 6). It stabilizes at a factor of 1.34 +/- 0.04.

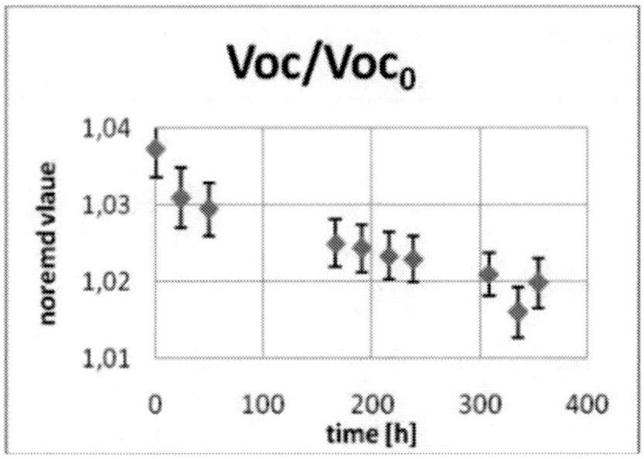

Figure 8. Open circuit voltage Voc of a-Si/μc-Si after stabilized at a degradation temperature at 50°C and during light-induced degradation at 30°C normed at its initial value (Voc_0)

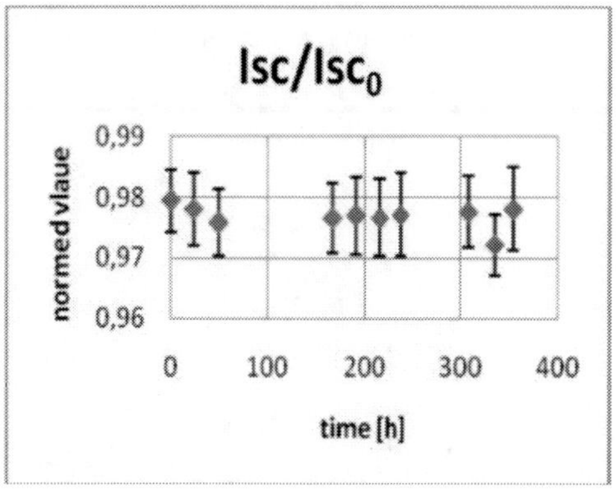

Figure 9. Short circuit current Isc of a-Si/μc-Si after stabilized at a degradation temperature at 50°C and during light-induced degradation at 30°C normed at its initial value (Isc_0).

This stabilized factor is higher than if the samples would have been degraded at a constant temperature of 30°C. The increasing effect of Rs was already observed under field test conditions [8].

The parallel resistance Rp decreases. After a small increasing in the first hours (figure 7), Rp decreases with a high fluctuation.

Figure 8 shows Voc during the procedure of degradation. Before the samples were degraded at 30°C, Voc increases at the degradation of 50°C. Voc decreases at a degradation temperature of 30°C.

As already shown for a-Si solar cells, Isc decreases with lower temperatures for a-Si/μc-Si solar cells.

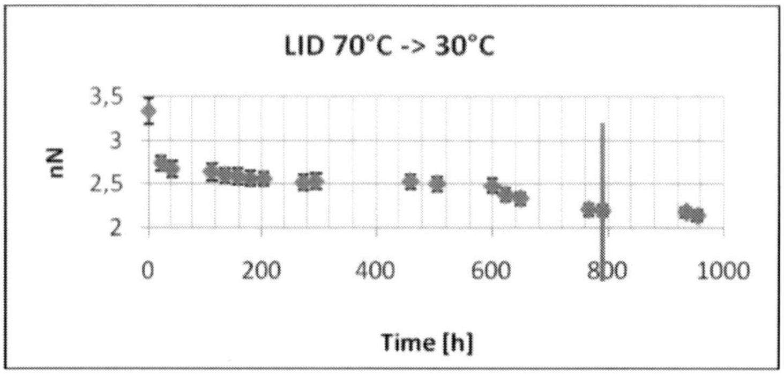

Figure 10. Diode factor n for a tandem solar module (N=2). After 600h the degradation temperature was decereased from 70°C to 30°C

The diode factor decreases stronger with lower degradation temperatures. The diode factor is calculated for one tandem solar cell, composed of one a-Si and one μc-Si solar cell. The diode factor stabilizes at about 200h of degradation, independent of the degradation temperature.

3.3. Degradation of a-Si/μc-Si solar cells at humid conditions

Samples of non- encapsulated a-Si/μc-Si solar cells were exposed to dark and high humidity conditions. We observed a decreasing efficiency (figure 11). This decreasing effect is lower than the decreasing effect described by the Staebler-Wronski effect. It is stabilized in less than 200h. The humidity air does not have any contact to the silicon layers of the solar cell. Because the solar modules are not encapsuled, the zinc oxide is in direct contact to air and its high humidity. By the absorption of humidity by the TCO the serial resistance increases [9]. By this increasing of Rs the ohmic loses in the TCO increases. By this increasing loses the efficiency of the solar cells decreases.

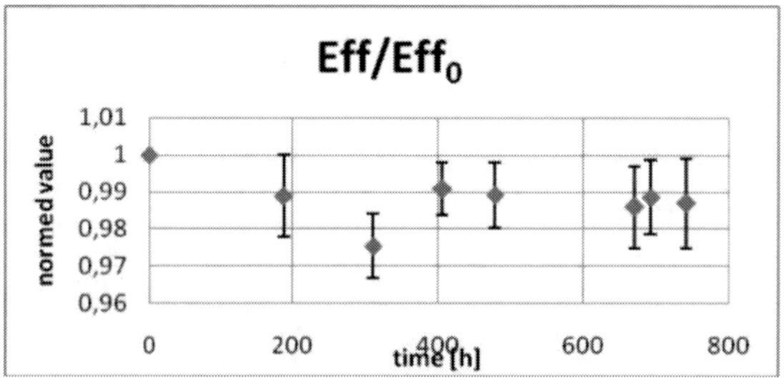

Figure 11. Efficiency (Eff) degraded at humidity conditions normed at its initial value

4. Conclusions

The investigation of light induced degradation of micromorph solar modules under different degradation temperature shows, that the serial resistance in the equivalent circuit increases at lower temperatures because of the light-induced degeneration of the Staebler Wronski effect. The average degradation temperature plays a central role for the equivalent circuit of micromorphous solar modules and its light-induced degradation behavior.

Also it was shown, that humidity air has a decreasing effect of non-encapsulated solar cells: The humidity affects the zinc-oxide and decreases the efficiency of the solar cell: By increasing the serial resistance the ohmic loses is higher in the zinc oxides. Encapsulating the solar cells with a back glass can decrease the decreasing effect.

5. Acknowledgments

We acknowledge support from BMBF through Contact No. 03FH008I2 and the support from Ministry for Innovation, Science and Research of North Rhine-Westphalia.

References

[1] Staebler D and Wronski C R 1977 *Appl. Phys. Lett.* **31** 292
[2] Weicht J A, Hamelmann F U and Behrens G 2013 *Proc .EU PVSEC 28th (Paris, France 30. September – 4 November 2013)* 2542
[3] Quaschning V 1966 *Simulation der Abschattungsverluste bei solarelektrischen Systemen* (Berlin, Köster)
[4] Chan D S and Phang J C 1987 *IEEE Transact. Electron Devices* **34** 286
[5] Kołodziej A 2004 *Opto-electronics Review***12** 21
[6] Crandall R S 1991 *Phys. Review B* **43** 4057
[7] Meillaud F, Shah A, Droz C, Vallat-Sauvain E and Miazza C 2006 *Solar Ener. Mater. Solar Cells* **90** 2952
[8] Weicht J A, Hamelmann F U, A. Domnik and Behrens G 2014 *Proc. 40th IEEE PVSC* (Denver, Colorado, USA 8-13 June 2014)
[9] Yaklin M A, Schneider D A, Norman K, Granata J E and Staiger C L 2010 *Proc. 40th IEEE PVSC (Honolulu, Hawaii, USA 20-25 June 2010)*

Optical and structural properties of electrochemically deposited ZnO nanorod arrays suitable for improvement of the light harvesting in thin film solar cells

M Petrov[1], K Lovchinov[1], M Mews[2], C Leendertz[2] and D Dimova-Malinovska[1]

[1]Central Laboratory of Solar Energy and New Energy Sources, Bulgarian Academy of Sciences, 72 "Tzarigradzko Chaussee" blvd., 1784 Sofia, Bulgaria
[2]Helmholtz-Zentrum Berlin für Materialien und Energie, Institut Silizium Photovoltaik, Kekuléstr.5, 12489 Berlin, Germany

E-mail: petrov80@abv.bg

Abstract. The results of study of the optical and structural properties of ZnO nanorods (NR) arrays electrochemically deposited on two type substrates - the ITO surface on the front side of Si heterojunction (SHJ) solar cells and on stainless steel plate used for formation of a-Si:H thin film solar cells, are reported. The surface morphology of the NS arrays is examined by Scanning Electron Microscopy and AFM. The spectra of specular diffused and total reflection, and haze ratio in reflectance are compared before and after deposition of the ZnO NR arrays. In the case of deposition on ITO surface of SHJ solar cells the values of the direct and diffused reflection of the ZnO NR array decrease demonstrating good antireflection properties. Deposition of ZnO NS arrays on stainless steel plates leads to increasing the values of the diffused reflection and the total reflectance. Possible application of ZnO NS structures for the processing of advanced Si based solar cells for increasing light harvesting is discussed.

1. Introduction

Increasing the efficiency and decreasing the production cost of terrestrial photovoltaic devices is important for the further development and widespread of the photovoltaic technologies. Solar cells made applying nanostructured and new emerging materials are being investigated to address this challenge. One of the ways to get higher efficiency of solar cells is to increase the optical absorption by light trapping. This could be achieved by implementation of nanomaterials: i) for decreasing the reflection of the light from the front side of the solar cells [1-2] and, as well as, ii) for increasing the reflection and scattering of the passed through the solar cells light by three dimensional (3D) back side surface [3]. Additionally, the nanometer size objects have very large surface areas per unit volume which offers the possibility to form very large interfacial area resulting in increasing the charge carrier collection, the generated current and the efficiency of the solar cells due to the 3D geometry of vertical aligned nanostructured features as nanorods and nanowiskers [4, 5].

ZnO is material widely used for thin film solar cells. It can be grown in the large nanostructured arrays by a low cost method as electrochemical deposition [6, 7].

This work reports the results of study of the optical and structural properties of ZnO nanostructures on two type of substrates – on the ITO on the front side of the Si heterojunction (SHJ) solar cells - Ag grid/ITO/a-Si/c-Si/Al [8] and on the stainless steel plates to be applied as substrates for thin film a-Si:H solar cell deposition [9]. The obtained changes in the optical spectra of specular and total reflectance, and diffused reflection are discussed regarding the possible applications of the nanostructured ZnO arrays.

2. Experimental

ZnO nanostructured films were deposited on the front side of the heterojunction solar cells by an electrochemical process from aqueous solution of $ZnCl_2$ (5. 10^{-4} M) and KCl (1 M) with pH≈5.0 at 90^0C for for1 min - in the case of growing on the ITO face of SHJ; and at 60^0C for 30min – in the case of deposition on stainless steel plate using a three-electrode electrochemical cell with SCE as a reference electrode. As a working electrode SHJ solar cells and stainless steel plates were used. Spectrally pure graphite rod electrode was applied as anode. The electrolyte was agitated by magnetic stirrer.

Deposited zinc oxide layer could be suggested to be a product of interaction of OH^- and Zn^{2+} in aqueous solution on the surface of the electrode (equation 1). Since the zinc hydroxide is unstable, it does dehydrate into ZnO [10]:

$$Zn^{2+} + 2OH^- \rightarrow Zn(OH)_2 \rightarrow ZnO + H_2O \qquad (1)$$

The process was carried out controlling the redox potential of the high power potentiostat system WENKING HP 96. The deposition potential was kept at -1000 mV (vs. SCE). The oxygen content in solution was permanently measured and the redox potential was kept in the range 300-400 mV (vs. SCE), oxygen is provided in the system by supplement of H_2O_2.

The optical spectra (specular reflection, diffuse reflection) of the deposited arrays were measured by a spectrophotometer Shimadzu UV-3600 in the range of 300 - 1200 nm employing a 60 mm integrated sphere. The surface morphology studies were performed by a Scanning Electron Microscope (SEM) Philips 515 and by AFM apparatus NT-MDT Solver 47 Pro system in a "semi-contact" (taping) mode.

3. Results and discussion

3.1 ZnO nanostructures (NS) deposited on ITO front contact of SHJ solar cells

The ZnO nanorod arrays are electrochemically deposited on the front side (ITO layer) of the heterojunction solar cells - Ag grid/ITO/a-Si/c-Si/Al. In this case the deposition time is 1 min. As we reported earlier the XRD diffraction study of the deposited ZnO by this technique reveals the existence of ZnO with hexagonal wurtzite structure [11]. The SEM micrographs with different magnifications of the deposited ZnO nanostructured arrays are demonstrated in figure 1.

a b

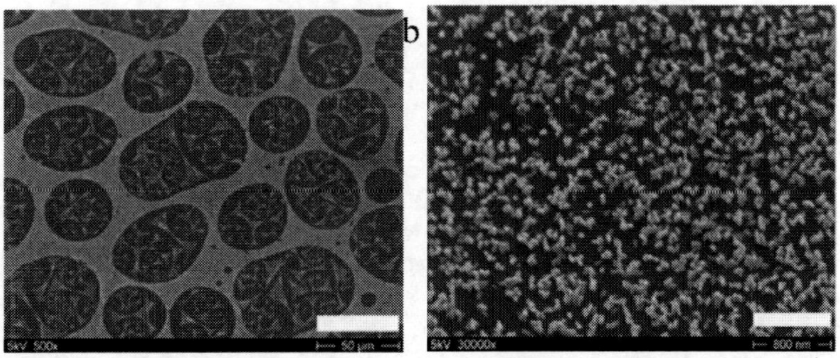

Figure 1. SEM of ZnO NS deposited on ITO front side of SHJ solar cell with different magnification. The markers correspond to 50 μm (a) and 800 nm (b).

The grown ZnO consists of nanograins which are distributed very inhomogeneous on the surface of the substrate – in the less dense large spot-like areas sized of several tens of μm^2 with rarely located grains are seen (figure 1a). It is seen clearly that on the surface of the denser areas the grains have size of about 60-80 nm and the distance between them is in the range 50 - 400 nm (figure 1b).

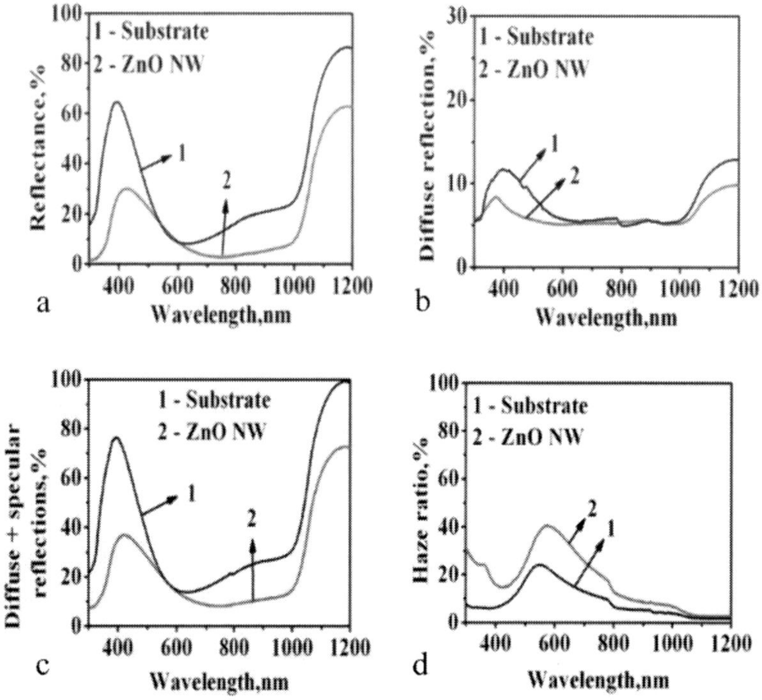

Figure 2 Spectra of reflectance (a), diffused reflection (b), total reflection ($R + R_{dif}$) (c) and haze ratio in reflectance (d) of ZnO NS array layer deposited on front side of SHJ solar cell. The corresponding spectra of the substrate before ZnO NS deposition are shown, as well, for comparison.

Figure 2 shows the spectra of specular reflectance (R_{spec}), diffused reflection (R_{dif}), total reflection ($R_{tot} = R_{dif} + R_{spec}$) and haze in reflectance (H_R) of the ZnO NR arrays grown on the front side of the SHJ solar cell. The haze ratio in reflectance is defined as a ratio of the diffused to total reflection:

$$H_R = R_{dif}/(R_{dif} + R_{spec}) \quad (2)$$

The values of reflectance and total reflection decrease after deposition of ZnO NR arrays. In the spectra of diffused reflection decreasing is observed in the wave length ranges: 350-650 nm and 1000-1200 nm.

Decreasing the values of the specular (figure 2 a) and total reflection (figure 2 c) and increasing the haze ratio in reflection (figure 2d) are seen in almost the whole spectral region. It should be noted that the maximal drop of reflection in the whole spectral region under investigation is observed for the sample with grain size of ZnO NRs of 60-80 nm (figure 1 b) deposited for 1 min on the front side of the SHJ solar cell compare to the other samples where the deposition time is longer as reported earlier [12]. The ZnO NRs arrays studied in this work demonstrate good antireflection (AR) properties. The minimal values of reflectance of about 2.5- 3% are observed in the larger spectral range 650-850 nm (figure 2a) than to the results showing minimum value of reflectance – 3-4 % only at 667 nm as reported by other authors for the ZnO NRs arrays with 50-60 nm diameter size of the NR obtained by solution grown on poly-Si substrates [13]. The study of the I-V characteristics of SHJ solar cells before and after deposition of ZnO NR arrays for 1 min shows improvement in the values of the short circuit current by 3.371 mA/cm^2 and of the efficiency by 1.6%, which correspond to relative increasing of 12.6% and 14.2%, respectively, after deposition of ZnO NR arrays. This is due to the decreasing of reflection from the front side of the solar cell and increased light harvesting earlier reported by us [12], which values are higher than those shown in [13, 14].

3.2 ZnO nanostructures (NS) deposited on stainless steel plates

Figure 3 demonstrates the SEM micrograph and AFM picture of ZnO nanorod arrays deposited on stainless steel plate. The hexagonal ZnO nanorods with size about 100-500 nm are grown almost perpendicularly to the substrate.

a

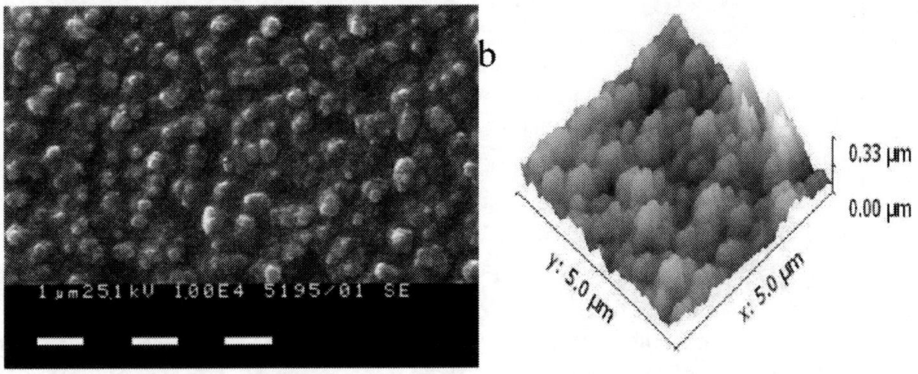

Figure 3. SEM micrograph (a) and AFM picture (b) of the ZnO layer electrochemically deposited on stainless steel plate.

They are relatively homogeneously grown on the whole surface area of the substrate. The ZnO NRs have different height with maximal size of about 300 nm (figure 3b). The estimated mean roots square roughness obtained from AFM measurement is 120 nm.

The spectra of specular reflectance, diffused reflection, total reflectance and haze ratio in reflectance of the ZnO NR arrays grown on stainless steel plate are shown in figure 4. It is seen that after deposition of ZnO nanorods array the values of reflectance increase for the wave length > 600 nm and diffused reflection and total reflectance values increase in the visible range 500-800 nm.

The haze ratio in reflectance of the ZnO nanorod array is higher compared to the values of the substrate in the spectral range 400 - 650 nm where the values of the specular reflection are lower than those of the substrate. The higher values of the diffused reflection and of the total reflectance after ZnO NRs growth on the stainless steel substrate could be explained by the optimal for maximal absorption of the light diameter size of the ZnO nanorods and surface roughness.

As is reported in [2] the 3D approach for modeling of the performance of the pin a-Si:H solar cells with AR ZnO NR arrays shows that the values of the nanorods diameter size of 150 nm and the surface roughness of 120 nm are optimal for the maximal absorption of the light by the rough surface. The size of the diameter of ZnO NRs grown on stainless steel plate, object of investigation in our work, is 100-500 nm and the surface roughness is 120 nm, which values are close to the optimal theoretical ones calculated in reference [2].

Journal of Physics: Conference Series **559** (2014) 012018 doi:10.1088/1742-6596/559/1/012018

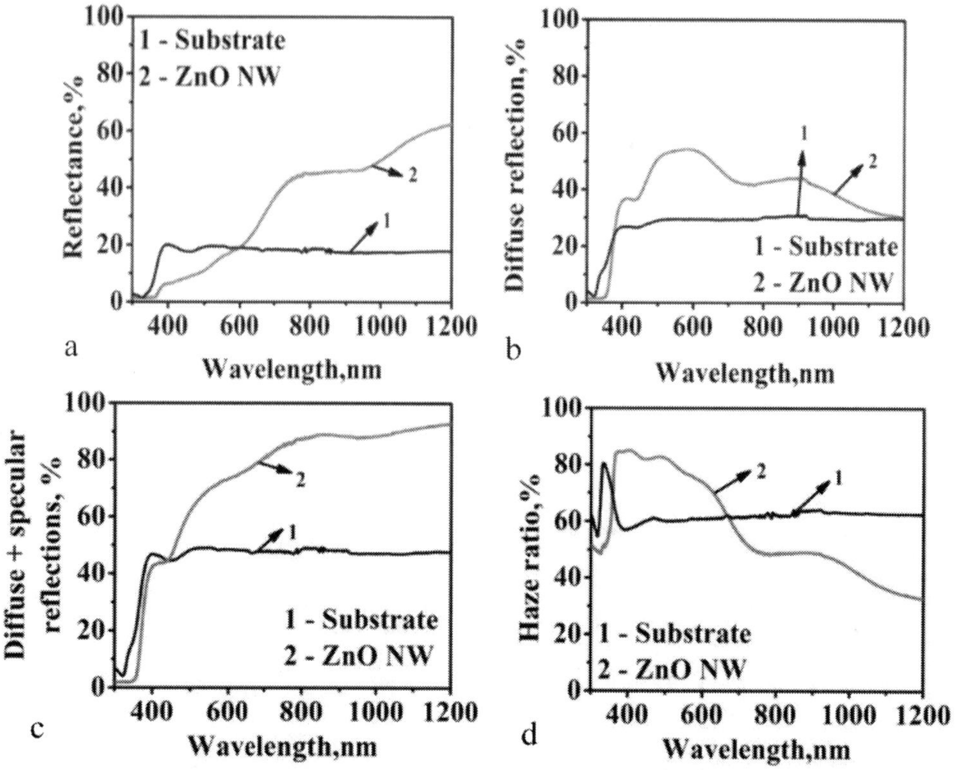

Figure 4. Spectra of reflectance (a), diffused reflection (b) and haze in reflection (c) of ZnO NS layers electrochemically deposited on stainless steel substrates. For comparison the corresponding spectra of the seeding stack structures (curves 1) are presented, as well.

Stainless steel plates with ZnO nanorod arrays are used as substrates for n-i-p a-Si thin film solar cells with ITO and Ag grid as front electrode, and Ag/ZnO:Al back reflector deposited on the substrate stainless steel/ZnO NRs arrays [9]. The preliminary results demonstrate that the J_{sc} of the solar cell with ZnO NRs is 2.4 mA/cm^2 higher than that of the solar cell without ZnO nanorods. It has to be notice that the comparison to the cell with nano-imprinted replica of Asahi-U texture, the J_{sc} gain is 1.4 mA/cm^2 [9]. These improvements correspond to relative increasing in the short circuit current and the efficiency by about 20% compare to the solar cells grown on the SS substrate without ZnO nanorods and about 10% compare to the reference cell deposited on SS substrate with nano-imprinted replica of Asahi-U texture at the rear side [9]. The improvement of the solar cells parameters can be explained by the increased diffused reflection and total reflectance of the substrate with ZnO NRs leading to higher light harvesting due to the suitable three-dimensional geometry of the nanorod arrays. Additionally, the larger effective area of the thin film solar cell deposited on the substrate with ZnO NRs arrays results in increasing the light trapping as well.

Conclusion

In this work the surface morphology and the optical properties of electrochemically deposited ZnO nanorod arrays on ITO on the front side of the SHJ solar cells and on the stainless steel plates applied as substrate for deposition a-Si:H thin film solar cells. The ZnO nanostructures arrays consist of nanorods with diameter size of 60-80 nm in the case of growing on the ITO front side of the SHJ solar cells and of about 100-200 nm when they are grown on stainless steel plate. Decreasing of the reflectance, diffused and total reflection after deposition of the ZnO nanorods on the front side of the

SHJ solar cells is observed. Due to the good anti-reflection properties of the deposited ZnO nanorod arrays the short circuit current and the efficiency of the SHJ solar cells are improved by 3.37 mA/cm^2 and of the efficiency by 1.6%, which correspond to relative increasing of 12.5% and 14.2 %, respectively. The ZnO nanorod arrays deposited on stainless steel substrates result in increasing of the diffused reflection and haze ratio in reflectance in the visible range of the spectra due to the appropriated size of the ZnO NRs and the roughness of the deposited nanostructured arrays used as substrate for deposition of a-Si:H solar cells. The relative improved by about 20% photocurrent and efficiency of thin film a-Si:H solar cells deposited on the stainless steel substrate with ZnO NRs compare to the reference cells are explained by increasing of light harvesting due to the appropriate size of the nanorods in three-dimensional arrays.

Acknowledgements
The work has been funded by the 7 European FP - project NanoPV № 24331 and the participation of M. Petrov and K. Lovchinov at the Workshop on "Transition Metal Oxide Thin Films-Functional Layers in Smart Windows and Water Splitting Devices: Technology and Optoelectronic Properties", 4-6 September, 2014, Varna, Bulgaria, has been supported by the INERA REGPOT Project of the ISSP- BAS and by Project BG051PO001-3.3.07-002 of EU and ESF.

References
[1] Battaglia C, Escarre J, Söderström K, Charriere M, Despeisse M, Haug F-J and Ballif C 2011 *Nature Phot.* **5** 535
[2] Geisendörfer S, Vehse M, Voss T, Richters J-P, Hanke B, von Meydel K and Agert C 2013 *Sol. Ener. Mater. Solar Cells* **111** 153
[3] Lee B G, Stradins P, Young D L, Alberi K, Chuang T K and Couillard J G *2011 Appl. Phys. Lett.* **99** 064101
[4] Vanecek M, Neykova N, Babchenko O, Purkart A, Poruba A, Remes Z, Holovsky J, Hruska K, Meier J and Kroll U 2010 *Procc 28 EUPVSEC 6-10 Sept.* Valencia Spain 2763
[5] Kuang Y, van der Werf K H M, Houweling Z S and Schropp R E I 2011 *Appl. Phys. Lett.* **98** 1131
[6] Kopenkamp R, Boedecker K, Lux-Steiner MC, Poschenrieder M, Zenia F, Levy Clement C and Wagner S 2000 *App. Phys. Lett.* **77** 2575
[7] Dimova-Malinovska D, Lovchinov K, Ganchev M, Angelov O, Graff JS and Ulyashin A 2013 *Phys Stat Sol* **A 210** 737
[8] Mews M, Schulze T S, Mingirulli N and Korte L 2013 *Energy Procedia* **38** 855
[9] Zhang D, Petrov M, Dörenkämper M, Bakkea K, Soppe W, Dimova-Malinovska D and Schropp R 2014 *29 EUPVSEC, 26-30 Sept.*, Amsterdam, Nederland - submitted
[10] Lide D R and Frederikse H P R 1996 *Handbook of Chemistry and Physics* 77th ed. Boca Raton FL CRC Press
[11] Lovchinov K, Ganchev M, Petrov M, Nichev H, Rachkova A, Angelov O, Mikly V and Dimova-Malinovska D 2013 *Phys Status Solidi A* **210** 743
[12] Petrov M, Lovchinov K, Ganchev M, Dimova-Malinovska D and Leendertz C 2013 *Proc. Scientific Studies Plovdiv University* **38** 4
[13] Chen J Y and Sun K W 2010 *Solar Ener.Mater. Solar Cells* **94** 930
[14] Yu X, Wang D, Lei D, Li G and Yang D 2012 *Nanoscale Res. Lett.* **7** 306

Journal of Physics: Conference Series **559** (2014) 012019 doi:10.1088/1742-6596/559/1/012019

Biogenic oxides from neutrophilic iron bacteria and possibilities for application in the nanotechnology

R Angelova[1,2], B Blagoev[2], L Slavov[2], M Iliev[1], V Groudeva[1] and I Nedkov[2]

[1] Faculty of Biology, Sofia University, 8 Dragan Tsankov blvd., 1164 Sofia, Bulgaria
[2] Institute of Electronics, Bulgarian Academy of Sciences, 72 Tzarigradsko Chaussee blvd., 1784 Sofia, Bulgaria

E-mail: raly_angelova@abv.bg

Abstract. The aim of this study is to obtain and characterize the ferric oxides/(oxy)hydroxides formed after cultivation of bacteria under laboratory conditions. The pure cultures of these bacteria isolated from natural habitats are identified by the methods of classical and molecular taxonomy as strains of the *Leptothrix* genus. Adler (AM) and Silicon iron glucose peptone (SIGP) media are the most appropriate ones for obtaining the iron oxides. The characterization of the oxides and sheaths is performed by different physical methods. The sheaths are formed in a SIGP medium. Light micrograph images and SEM revealed the average size and diameter of the sheaths. The XRD measurements showed the composition of the oxides obtained, as well as the average size of the iron particles (up to 30 nm). The TEM micrographs showed the shape of the biogenic nanoparticles, while the magnetic measurements demonstrated the superparamagnetic character of the magnetic part of the biomaterials. The new biogenic materials are promising for application in magneto electronic for building biosensors.

1. Introduction

The field of nanosized biogenic iron oxides is truly multidisciplinary. The objects of this investigation are the iron sub-products obtained by laboratory cultivation of neutrophilic iron bacteria. These bacteria are a large physiological group with different taxonomic status but united by their ability to oxidize Fe(II) at neutral pH [1] in the appropriate nutrient media. As a result of their metabolism, they form biogenic iron oxides/(oxy)hydroxides accumulated on the nanosized powder and organic sheaths. In the tubular structures, the iron oxide formations are magnetic single-domains discretely dispersed into an organic matrix. The superparamagnetic iron oxides domains with appropriate surface chemistry are promising materials for magneto electronic applications as biosensors. On the other hand, this structure absorbs light in the IR spectrum and can self-assemble in a magnetic field and replace the supra-molecular assembly of light sensitive polymers thus opening up possibilities of constructing molecular-level devices.

The bacteria-mediated formation of iron-containing tubular structures and their magnetic properties are currently poorly understood. It is also not clear why many of the strains do not form sheaths during cultivation under laboratory conditions [2]. This investigation is focused on the iron sub-products containing powder and sheaths obtained by two different types of cultivation, which are developed in the Department of General and Industrial Microbiology, Faculty of Biology, Sofia University [3]. The neutrophilic iron bacteria from the *Leptothrix* genus isolated from Vitosha Mountain (nearby Sofia, Bulgaria) are identified by the methods of classical [1] and molecular taxonomy.

2. Materials and methods

The cultivation of the bacteria of the *Leptothrix* genus is carried out in the following two different nutrient media:

(AM) - Adler's medium – modification: Sodium lactate - 40.0 mg; Yeast extract - 1.0 g; Ascorbic acid - 0.1 g; $MgSO_4.7H_2O$ - 0.2 g; K_2HPO_4 - 0.01 g; $(NH_4)_2Fe(SO_4)_2{\cdot}6H_2O$ - 0.01 g; dH_2O - 1000 ml; pH 7.0.

(SIGP) - Silicon iron glucose peptone: 1 g glucose, 1 g Bacto peptone (BD, France), 0.2 g $Na_2SiO_3\_9H_2O$, 0.044 g $CaCl_2\_2H_2O$, 0.041 g $MgSO_4\_7H2O$, 0.076 g $Na_2HPO_4\_12H_2O$, 0.02 g $KH_2PO_4\_2H_2O$, 2.838 g HEPES (N-2-hydroxyethelpiperazine-N'-2-ethanesulfonic acid), and 0.05 mM $FeSO_4$ in 1000 ml of distilled water (pH 7.0).

As additional source of iron, iron cuttings are supplemented to both media. The cultivation is carried out at 20 °C and a pH of 7.0. The changes in the pH and the content of Fe(II) in the cultivation vessels are monitored periodically. During the whole period of cultivation, a microscopic observation is performed of the cultural liquid from all samples.

The analysis of the iron oxides/(oxy) hydroxides is performed by X-ray diffraction (XRD) which is used to determine the phase composition of the biogenic samples. Scanning electron microscopy (SEM) and transmission electron microscopy (TEM) are used to study the bacteria ability to produce sheaths in such an environment. The magnetic properties of the samples are measured by a physical properties measurement system (PPMS, Quantum Design). The real and imaginary parts of the differential magnetization MAC are measured from zero-field-cooled (ZFC) 10 K to 300 K in an AC 10-Oe magnetic field at a frequency of 1 kHz. Hysteresis loops at 300 K up to 10 kOe are also recorded of both types of samples.

All measurements are conducted on samples taken after 40 days of cultivation in order to determine the material type. The cultural liquid of the samples is filtered and dried.

3. Results and discussions

The bacteria formed the typical sheaths (figure 1a) during the cultivation on SIGP medium, while on AM medium the bacterial cells are observed only (figure 1b). The formation of the sheaths started after a seven-day cultivation period. Their average diameter is in the range of 500 – 1000 nm, with the length reaching approximately 7 μm. The structures disintegrated completely approximately after 90 days since cultivation.

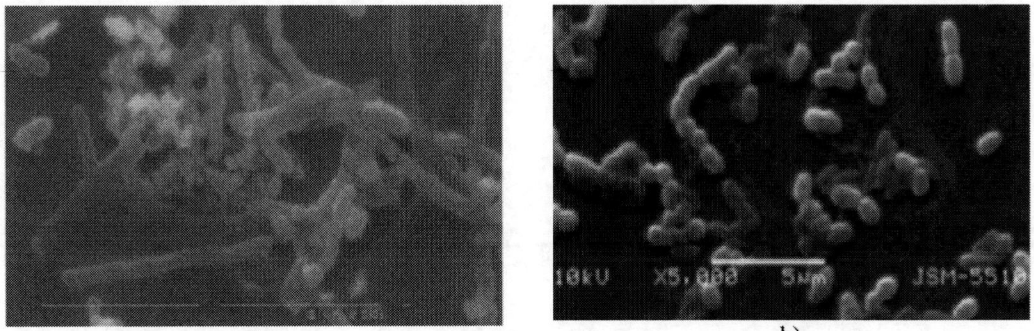

a) b)

Figure 1. SEM of sample from: a) a SIGP medium, ×5000; and b) an AM medium, ×5000.

The XRD spectra of the biogenic material obtained showed the type of the oxides/(oxy)hydroxides, their percentage ratio and the particle size [4]. The XRD data for the AM sample revealed three phases – lepidocrocite (γ-FeOOH), goethite (α-FeOOH) and magnetite (Fe_3O_4), with dimensions of the particles up to 30 nm. In the material collected from the SIGP-cultivated *Leptothrix* spp., the XDR spectra revealed a single-phase composition – lepidocrocite (γ-FeOOH) with an average diameter of the particles of 8 nm [5]. The tubular samples spectrum revealed the existence of an organic matrix.

As can be seen in figure 2 and 3, the magnetization of sample AM is one order of magnitude higher than that of SIGP. Such a result is to be expected because the SIGP sample consisted of 100 % antiferromagnetic (AF) lepidocrocite, while there is ~ 20 % of magnetite present in the AM sample. Nevertheless, the imaginary part of the differential magnetization corresponding to magnetic losses increased rapidly at low temperatures and reached a sharp maximum at about 30 K for both samples. Moreover, the differential magnetization real part in the temperature range 10 – 50 K increased nonlinearly and also rapidly. Such an identical behavior of the differential magnetization could be due to a temperature transition in lepidocrocite. Under these measurement conditions, the peaks of the real and imaginary parts of the differential magnetization of magnetite are at about 215 K and 125 K, respectively [6]. At higher temperatures (above 50 K), both magnetization real parts became linear; that of AM increased slightly up to 300 K, while for SIGP, saturation could be seen up to 160 K, followed by a linear increase up to 300 K (figure 2). These differences at higher temperatures are due to the different compositions of the samples. The hysteresis loops at 300 K until 10 kOe seemed to be the same except for the strength of magnetic response.

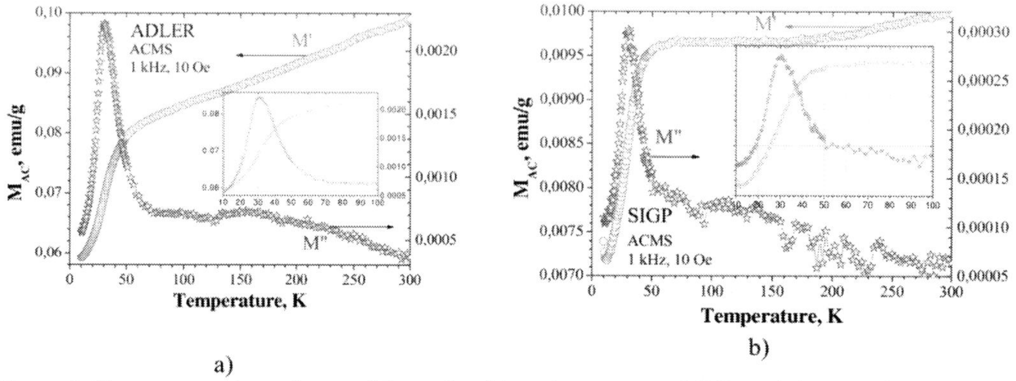

Figure 2. Temperature dependence of the real and imaginary parts of differential magnetization MAC of the sample a) AM and b) SIGP at an AC magnetic field 1 kHz, 10 Oe. The circles show the real part, the asterisks, the imaginary part. The insets show the magnetization around the temperature transition.

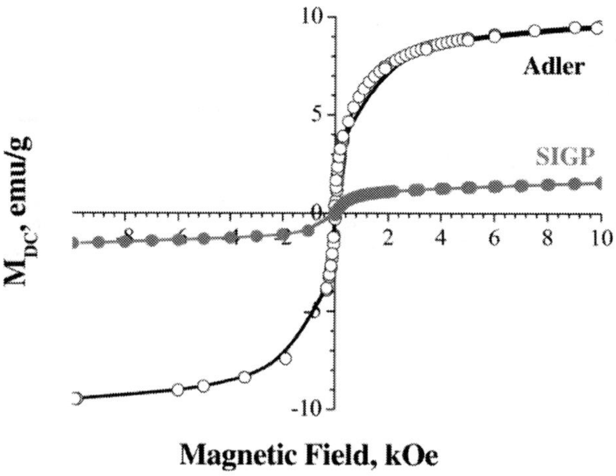

Figure 3. The hysteresis curve $M(H)$ at 300 K up to 10 kOe for samples Adler (AM) and SIGP..

5. Conclusions

Two iron-containing biogenic compounds with different magnetic nature are compared. One of the samples consisted of 100 % AF lepidocrocite, while the other sample is a mixture of AF (lepidocrocite and goethite) and ferrimagnetic (magnetite) components. The magnetic hysteresis loops at 300 K up to 10 kOe and the differential magnetization vs. temperature up to 50 K had the same behavior as AF lepidocrocite. The temperature dependence of the differential magnetization above 50 K of the two samples showed a different behavior. The magnetization of the magnetically mixed sample AM increased linearly and slightly until 300 K, while the magnetization of the AF SIGP sample saturated up to 160 K and increased linearly afterwards. Additional magnetic measurements are needed to understand fully the magnetic nature of these compounds.

Acknowledgements

This work was supported by the Bulgarian National Science Fund under project ДИД 02/38/2009, the project 165/04/2014 of Sofia University Science Fund, and INERA project of the Institute of Solid State Physics at Bulgarian Academy of Sciences.

References

[1] Mulder E G 1989 *"Sheathed bacteria" in Bergey's manual of systematic bacteriology* edited by Staley T et al. (The Williams & Wilkins Co., Baltimore)
[2] Emerson D and Chiorse W C 1992 *App. Environ. Microbiol.* **58** 4001
[3] Angelova R, Iliev M, Mitova M, Voynova V, Tasheva L, Slavov L, Groudeva V and Antonova-Nikolova S 2014 *J. Bio Sci. Biotech.* SE/ONLINE: 67-70
[4] Slavov L, Angelova R, Blagoev B, Kovacheva D, Abrashev M V, Iliev M, Gelev Ch, Groudeva V and Nedkov I 2014 Nanostructured Biogenic Iron Oxides Produced by Neutrophilic Iron-oxidizing Bacteria *GEOMICROBIOL J* (submitted)
[5] Angelova R, Nedkov I, De Grave E, Slavov L, Iliev M, Blagoev B, Kovacheva D and Groudeva V 2014 Magnetic Properties of Nanosized Iron-Containing Biogenic Materials after Cultivation of Bacteria from the Genus *Leptothrix* under Laboratory Conditions *AIP Materials* (submitted)
[6] Nedkov I, Slavov L, Blagoev B and Krezhov K 2013 *Bulg. J. Phys.* **40** 348

Electrochemically deposited ZnO nanostructure on ZnO:V seeding layer

K Lovchinov[1], M Petrov[1], V Mikli[2] and D Dimova-Malinovska[1]
[1]Central Laboratory of Solar Energy and New Energy Sources at the Bulgarian Academy of Sciences, blvd. Tsarigradsko chaussee 72, Sofia 1784, Bulgaria
[2]Materials Research Centre, Tallinn University of Technology,19086 Ehitajate tee 5, Tallinn, Estonia

E-mail: lov4@abv.bg

Abstract. ZnO nanostructured (NS) arrays are grown by electrochemical method on conductive seeding layers of ZnO doped with V (ZnO:V) deposited by magnetron sputtering on glass substrates. The surface morphology of the ZnO NS is studied by Scanning Electron Microscopy. The spectra of transmittance, diffused and specular reflectance are measured for the ZnO NS arrays deposited for 30 and 60 min. and are compared to the corresponding spectra of the substrate with seeding ZnO:V film. The values of diffused reflection are higher than those of the seeding layers due to the grown of nanostructured ZnO arrays. Increasing of the diffused reflection with the time of electrochemical growth is observed and is explained by increasing size of the grains. The results show the possibility for deposition of nanostructured ZnO arrays on low cost ZnO:V seeding layers with large effective surface area that could be applied in thin film solar cells for increasing light harvesting.

1. Introduction

Nanostructured semiconductor materials have attracted the interest of the researchers due to their potential for different innovative applications, e.g. in solar cells [1] as an n-type electrode [2], near UV-leasers [3], field effect transistors [4], gas sensors [5], etc.

Application of the nanomaterials in the architecture of thin film solar cells started with proposed by M. Green idea [6] for development of the third generation solar cells. The goal is the development of thin films low cost technology for solar cells and modules with higher efficiency. One way is the development of the preparation methods and study of properties of nanomaterials for applications in solar cells for improvement of the light harvesting and their efficiency.

ZnO is one of the most attractive semiconductor materials which can be produced in thin film with different nanostructured modifications [7, 8].

Different substrates have been used for growing ZnO nanostructured arrays such as ITO on glass, c-Si [8], multicrystalline Si [7], Cu plates [7].

In this work the results of study of the structural and optical properties of ZnO nanostructured arrays grown by electrochemical method on seeding conductive ZnO:V thin films are presented.

2. Experimental

Thin ZnO:V films were prepared by R.F. magnetron co-sputtering of ZnO target (10 cm diameter) with pieces of vanadium chips on in the maximum erosion zone of its surface in Ar atmosphere at a pressure of 0.5 Pa and r.f. power of 150W as described in [9,10].

The films were deposited on corning glass without heating of the substrates. The thickness of the ZnO:V films used as seeding layers for electctrochemical growth of ZnO NS's was 600 nm. The concentration of V was 1.41 % and the sheet resistance - 26 Ω/cm^2.

ZnO nanostructured films were deposited on the ZnO:V seeding layer by electrochemical process from aqueous solution of $ZnCl_2$ ($5x10^{-4}$ M) and KCl (1 M) with pH≈5.0 at 70^0C for different deposition times: 30 and 60 min using a three-electrode electrochemical cell with SCE as a reference electrode.

Spectrally pure graphite rod electrode was applied as anode. The electrolyte was agitated by magnetic stirrer. The process was carried out controlling the redox potential of the high power potentiostat system WENKING HP 96. The deposition potential was kept at -1000 mV (vs. SCE). Good quality ZnO films were obtained at a redox potential within the range between +0.300 and +0.400 V (vs. SCE). At a redox potentials higher than +0.4 V (vs. SCE) Zinc peroxide (ZnO_2) was formed on the samples with bad adhesion to the substrate. The oxygen content in solution was permanently measured, oxygen is provided in the system by supplement of H_2O_2.

The study of the surface morphology of the deposited ZnO nanostructured arrays was performed by a Scanning Electron Microscope (SEM) Philips 515. The optical spectra (transmittance, specular reflectance, diffuse reflection) of the deposited arrays were measured by a spectrophotometer Shimadzu UV-3600 in the range of 320 - 1800 nm employing a 60 mm integrated sphere.

3. Results and discussion

Figure 1 shows SEM surface and cross section images of the ZnO:V seeding layer. The SEM pictures show a dense column structure of the films with diameter of the column about size 30 nm.

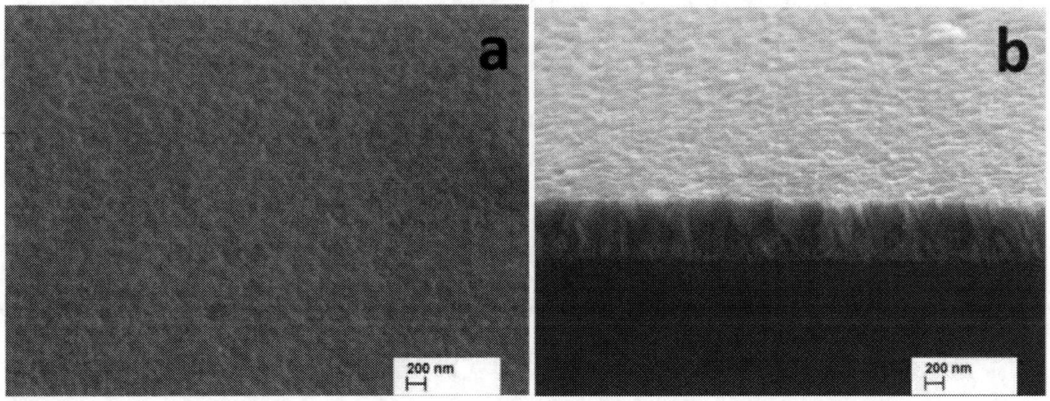

Figure 1. SEM of the seeding ZnO:V layer : plane (a) and cross section view (b).

The surface morphology of the ZnO structures grown electrochemically for 30 and 60 min on these seeding layers is presented on figure 2.

It can be seen from figure 2 that ZnO structures deposited for 30 min consist of: grains with diameters of about 1 – 2 μm grown between thin walls that look like whiskers. These walls have different lengths—from 1 to 10 μm and thickness of about 200 nm. An increase in the deposition time to 60 min leads to increasing the number of the grains with size about 1 μm, some of them coalesce between them and with the walls as a result larger grain areas of about 3-4 μm are formed. The height of the grain increases from about 0.5 μm to 1 μm with deposition time increasing till 60 min.

Figure 2. SEM pictures of the ZnO nanostructures grown on ZnO:V seeding layer for different deposition time (a, b and c)- 30 min, and (d, e and f) – 60 min; pictures (a, b, d and e) shown - surface view, and pictures (c) and (f) shown cross section. The markers in the pictures (a and d) correspond to 10 μm and in the pictures (b, c, e and f) correspond to 1 μm.

The spectra of transmittance, specular reflectance and diffused reflection of the ZnO nanostructured area layer are presented in figure 3 a, b, c, respectively. The spectra have been recorded in the optical range of 320 – 1800 nm. For comparison, the corresponding spectra of ZnO:V seeding layer are presented as well. The values of transmittance and specular reflectance of the samples with deposited ZnO nanostructure decrease in the visible and near IR range compare to the value of the substrate. However the values of diffused reflection are higher compare to value of the seeding layer and increase with the time of ZnO electrochemical deposition due to the more developed surface of the deposited ZnO nanostructured arrays – the grains become bigger.

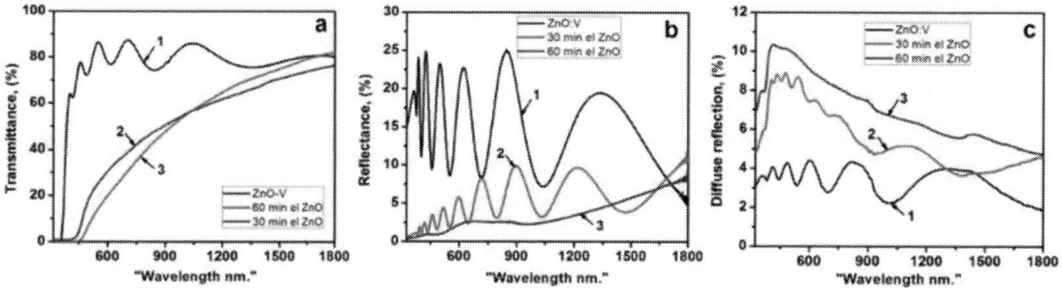

Figure 3. The spectra of transmittance (a), reflectance (b) and diffused reflection (c), of the ZnO nanostructured electrochemically deposited layer for different time: 30 min - curves 2 and 60 min – curves 3. For comparison the spectra of ZnO:V seeding layer are shown – curves 1.

Similar results have been reported for deposition of ZnO NS array on other substrate materials, such as multicrystalline Si [7, 11]. The layers grown on low cost ZnO:V substrate can be considered as supporting substrates for deposition of a-Si:H, poly-c-Si, organic etc. thin film based solar cells.

4. Conclusion

It is demonstrated for first time, that conductive ZnO:V thin films can be used as seeding layers for nanostructured ZnO arrays grown by electrochemical deposition. It is observed that ZnO nanostructures consist of grains and wiskers. An increase of the diffused reflection compared to that of the seeding ZnO:V layers is due to the nanostructures grown. With increasing of the time of deposition the nanograins increase in size leading to formation of more developed surface and increased values of the diffused reflection.

Although further optimization of such ZnO NRs based structures is required, the obtained results can be considered as a promising technological development towards low-temperature and cost effective technology for the processing of substrates in thin film based solar cells with improved light harvesting.

Acknowledgement

The work has been funded by the 7FP - project № 246331 NanoPV, and the participation of K.Lovchinov at the Wokshop on "Transition Metal Oxide Thin Films-Functional Layers in Smart Windows and Water Splitting Devices: Technology and Optoelectronic Properties", 4-6 September, 2014, Varna, Bulgaria has been supported by the INERA REGPOT Project of the Institute of Solid State Physics and by the Bulgarian Scientific Fund – project N 859/2012, Project BG051PO001-3.3.07-002 of EU and ESF. The authors thank assoc. prof. Dr. O. Angelov for the deposition of the ZnO:V seeding layer.

References

[1] Gregg B 2009 *MRS Bull.* **30** 20
[2] Gonzales-Valls I and Lira-Cantu M 2009 *Energy Environ. Sci.* **2** 12
[3] Service R 1997 *Science* **276** 895
[4] Park W, Kim J, Yi G, Bae M and Lee H 2004 *Appl. Phys. Lett.* **85** 505
[5] Lovchinov K, Petrov M, Ganchev M, Georgieva V, Nichev H, Georgieva B and Dimova-Malinovska D 2014 *J. Phys.: Conference Series* **514** 012023
[6] Green M A, *Physica E* 2002 **14** 65
[7] Dimova-Malinovska D, Lovchinov K, Ganchev M, Angelov O, Graff J S and Ulyashin A 2013 *Phys. Status Soidi.* **A 210** 737
[8] Guske J T, Brown J, Welsh A and Franzen S 2012 *Optics Express* **20** 23215
[9] Lovchinov K, Angelov O, Nichev H, Mikli V and Dimova-Malinovska D, 2012 *Energy Procedia* **10** 282
[10] Angelov O, Lovchinov K and Dimova-Malinovska D 2013 *Phys. Status Solidi C* **10** 709
[11] Lovchinov K, Ganchev M, Rachkova A, Nichev H, Angelov O, Mikli V and Dimova-Malinovska D 2013 *Phys. Stat. Solidi A* **210** 743

Electron beam induced changes in the reflectance of materials

D Lyutov, S Hadjiiski, K Kirilov, G Tsutsumanova, A Tsonev and S Russev
Department of Solid State Physics and Microelectronics, Faculty of Physics, Sofia University "St. Kliment Ohridski", J. Bouchier 5, Sofia, Bulgaria

E-mail: d_lyutov@phys.uni-sofia.bg

Abstract. In this work we examine the opportunity to combine the advantages of two well-known experimental techniques – ellipsometry and scanning electron microscopy (SEM). Ellipsometry has very high sensitivity in the direction normal to the sample's surface while the lateral resolution is limited by the width of the probe beam, while the scanning electron microscopy has better lateral, but limited in-depth resolution. The electron beam can induce local changes in the electron density and the temperature of the sample, which alters the reflectance coefficient and can be potentially detected by optical methods and ellipsometry in particular.

1. Introduction

The ellipsometry is contactless and non-destructive method for researching of solid state or liquid surfaces and thin films. The information about the sample is achieved from the change of the polarization state of the probe light beam. Ellipsometry has high sensitivity for the optical properties and the thickness of layers – it can be used for detection of sub-monolayers. [1] However this resolution is in direction normal to the sample's surface, while the resolution parallel to the sample's surface is much lesser – it is limited by the width of the probe light beam.

The scanning electron microscopy has high resolution in the plane of the sample's surface (10 – 30 nm). The detected signal can be secondary electrons, backscattered electrons, photons (cathodeluminescence), X-ray. [2] Every of these signals is consequence of interaction of the electrons with matter.

In recent years, the scanning electron microscopy is combined with other experimental techniques for more detailed research of the materials. This can be accomplished with using more channels of information from the interaction of the electrons with matter. One example is using acoustic signal generated by modulated electron beam and detected by piezoelectric sensor. This technique is called scanning electron acoustic microscopy (SEAM) and it is used for imaging of boundaries of different phases or ferroelectric domains.[3]

A part of the electron beam energy is converted to heat which increases the local temperature. The electron beam also injects free carriers in the sample. The carrier injection and local heating lead to local changes of the refractive index. [4-7] Here we examine the possibility to use this still unexploited channel of information. To achieve our aim we complete two tasks: to perform several simulations of the optical response due to the local heating and to conduct some preliminary experiments.

2. Simulations of optical response

To examine the possibility to combine the two techniques (ellipsometry and SEM) several simulations are made. The induced changes of the refractive index for many types of materials are very small. The thermooptical coefficient is $dn/dT \approx 10^{-5}$ and $dk/dT \approx 10^{-3}$ for metals [8]. and $dn/dT \approx 10^{-4}$ and $dk/dT \approx 10^{-5}$ for semiconductors [9, 10] and $dn/dT \approx 10^{-5}$ for dielectrics [11, 12]. The optical response depends heavily on the angle of incidence and the input polarization state and it is very important to find the optimal conditions for the largest optical response. For this purpose we simulate the changes of the ellipsometric angles ψ and Δ and the reflection coefficients R_p and R_s induced by a small change of the refractive index at different angles of incidence and polarization.

The derivatives $d\psi/dn$, $d\Delta/dn$, dR_p/dn and dR_s/dn are calculated for angles of incidence φ in the range $0.01° < \varphi_i < 89.00°$ with a step $\Delta\varphi_i = 0.1°$. First, each of the parameters Ψ_0, Δ_0, R_{p0}, R_{s0} is calculated for varying angle of incidence (R_{p0} and R_{s0} are calculated using the scattering matrix approach [13]). Then a little value – $dn = 0.001$ is added to the refractive index n and the four parameters Ψ_1, Δ_1, R_{p1}, R_{s1} are calculated again. This addition dn simulates the change induced by the temperature alternation. In the end the numerical derivatives are obtained:

$$\frac{d\psi}{dn} = \frac{\psi_1 - \psi_0}{dn},\ \frac{d\Delta}{dn} = \frac{\Delta_1 - \Delta_0}{dn},\ \frac{dR_p}{dn} = \frac{R_{p1} - R_{p0}}{dn},\ \frac{dR_s}{dn} = \frac{R_{s1} - R_{s0}}{dn}.$$

For samples with complex refractive index $N = n - ik$ the derivatives are also calculated for the extinction coefficient k. The calculations are made for light source with wavelength $\lambda = 632.8$ nm. The derivatives of R_p and R_s are calculated to estimate the optical response in thermoreflectance and photoreflectance methods, where the relative change in the reflectance $\Delta R/R$ is measured. The thermoreflectance methods have very high sensitivity. The smallest detectable change of the reflectivity $\Delta R/R$ is in range less than 10^{-6} (10^{-7} range).[12, 14]

The simulations are made for three sample groups: with real refractive index, with complex refractive index and a layer with real refractive index on a substrate with complex refractive index. In figure 1 are shown the results for glass sample (real refractive index).

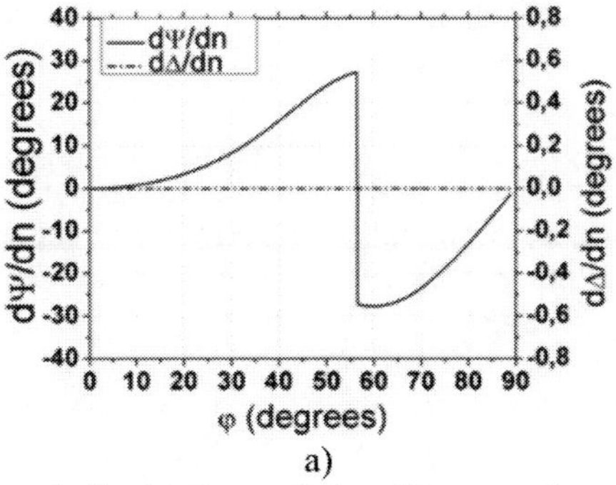

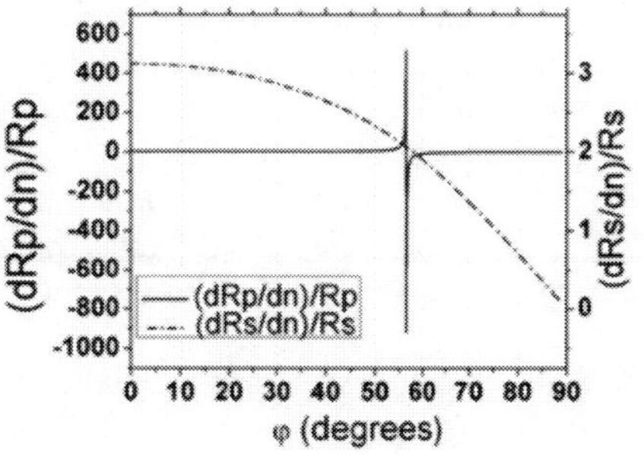

Figure 1. Derivatives of the ellipsometric parameters $d\psi/dn$, $d\Delta/dn$ a) and the reflection coefficients $(dR_p/dn)/R_p$, $(dR_s/dn)/R_s$ b) as function of the angle of incidence φ for glass sample $\lambda = 632.8$ nm; $N = 1.51$

It is obvious that the change of the reflection coefficient R_p is greatest close to the Brewster angle $\varphi = 57°$. For *s*- polarization the sensitivity is higher close to the angle of incidence 0° (normal incidence).

The simulations for the second group of samples (complex refractive index) in this case gold are shown in figure 2.

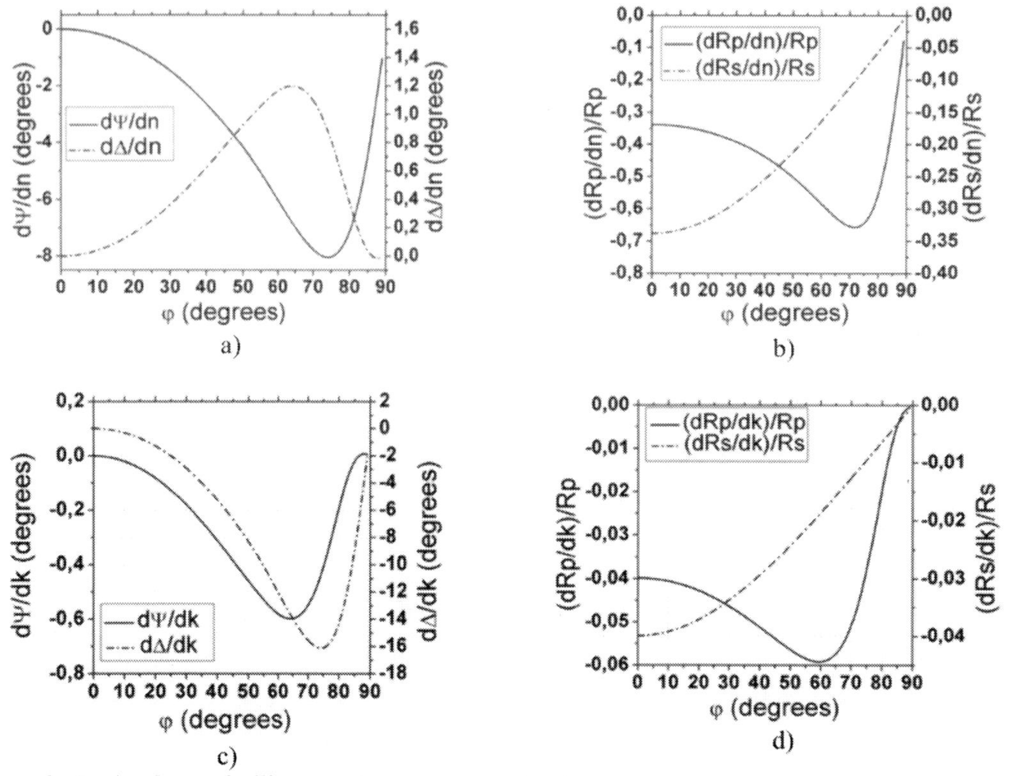

Figure 2. Derivatives of ellipsometric parameters *dψ/dn*, *dΔ/dn* a), *dψ/dk*, *dΔ/dk* c) and reflection coefficients $(dR_p/dn)/R_p$, $(dR_s/dn)/R_s$ b), $(dR_p/dk)/R_p$, $(dR_s/dk)/R_s$ d) as function of the angle of incidence for gold $\lambda = 632.8$ nm, $N = 0.21\text{-}3.272i$

The greatest changes in the ellipsometric parameters ψ and Δ are in the range $60° < \varphi < 80°$. The changes of the real part of the refractive index lead to greater changes in the reflection coefficients than the imaginary part. R_s changes mostly at angles close to 0°, while R_p changes mostly at angle $\varphi = 75°$ for alternation of real part n and at angle $\varphi = 60°$ for alternation of imaginary part k.

The third group of the model system is a layer with real refractive index on a substrate with complex refractive index. This simulation is more complex because there are six derivatives – four describing changes of the substrate and two describing changes of the layer. Representative results for a silicon dioxide layer (SiO_2, with thickness 5nm and 180 nm) on silicon (Si) substrate are shown in figure 3.

The simulations show that for layer thickness 5nm, the optimal angle of incidence for the ellipsometric angles ψ and Δ is $\varphi \approx 75°$. The useful range of incidence angles with acceptable optical response increases with increasing the thickness of the layer – see figure 6 c), d). The range where R_p and R_s have greater sensitivity also increases with thickness of the layer.

In order to estimate the detectable refractive index change induced by SEM irradiation, the power of the electron beam was calculated from the measured electron beam current at different accelerating voltages. The maximum power attainable in the SEM was 0.5 W. Using this value the effect of the electron beam heating onto different substrates was simulated. The results show that the temperature change is from several tens to several hundreds degrees.

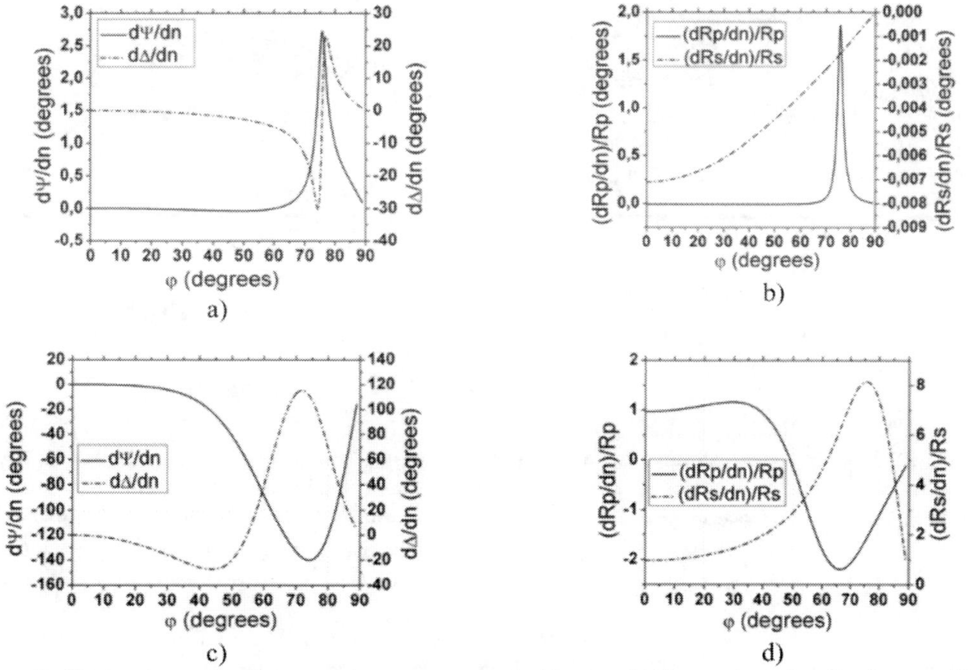

Figure 3. Derivatives of ellipsometric parameters $d\psi/dn_{SiO2}$, $d\Delta/dn_{SiO2}$ a) and reflection coefficients $(dR_p/dn)/R_p$, $(dR_s/dn)/R_s$ b) for layer with thickness 5nm as function of φ; $d\psi/dn_{SiO2}$, $d\Delta/dn_{SiO2}$ c) and reflection coefficients $(dR_p/dn_{SiO2})/R_p$, $(dR_s/dn_{SiO2})/R_s$ d) for layer thickness 180 nm as function of φ

On the other hand the uncertainty of the ellipsometric parameters is $d\psi = 0.02°$ and $d\Delta = 0.03°$. This is used for estimation of the smallest detectable changes in the refractive index, which are in the order of $\Delta n = 0.002$ and $\Delta k = 0.002$. These changes in the (complex) refractive index could be achieved by rising the temperature with approximately 20 K for semiconductors and 200 K for metals and dielectrics, which is achievable by electron beam heating.

3. Experiments

Two proof-of-concept experiments were conducted. The heating by the electron beam is modelled with heating by electric current (the first experiment) and focused laser beam (the second experiment).

In the first, gold layer deposited on glass (thickness d = 20 nm) is heated by rectangle electric signal (figure 4 a)). Its resistance is R = 76 Ω. The frequency and the duty cycle can be regulated. The maximum voltage amplitude is V_{pp} = 10 V and the maximum power is P = 1.3 W. The bifurcated optical fibre is used to illuminate the surface (semiconductor laser with λ = 532 nm) and to detect the reflected signal with a photo detector (Thorlabs PDA36A) – figure 4 a). The signal is fed to a lock-in amplifier with its reference channel connected to the output of the signal generator. The signal was detected with this set-up, but with very low signal to noise ratio. Strangely enough, a collateral result was far more stable detection of the acoustic signal generated by the heat modulation of the layer.

The second modelling experiment is similar to photoreflective methods. A semiconductor laser (λ = 532 nm) was used for modelling the electron beam heating and another semiconductor laser (λ = 659 nm) - for probe beam. They are focused by microscope at one point with objective with magnification x20 (figure 4 b)). The heating beam is modulated with mechanical chopper with frequency f = 723 Hz and its power is P = 130 mW. Interference filters are used for blocking the heating beam. Measurements are made for two samples Si and InP. The results are in table 1.

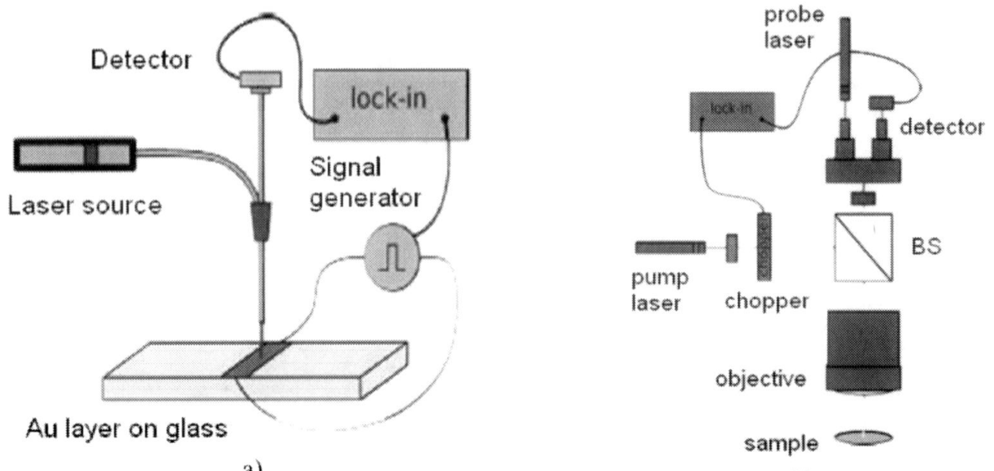

a) b)

Figure 4. a) Modelling experiment – modulated heating of gold layer (thickness 20 nm) with rectangle electric signal and detection of the changes of the reflectance b) Modelling experiment – heating with laser. A microscope is used to focus the heating and probe beam. The pump laser is modulated by a mechanical chopper. The probe laser is placed at the one ocular and the detector at the other.

The temperature change is estimated for the two experiments. The temperature rise is of the order of 1 K in the first experiment and of the order of 10 K in the second experiment.

The power of the electron beam is greater than that of the laser beam and the electron beam can be focused on smaller area with diameter 1 μm or less, while the laser beam can be focused on area with diameter approximately 10 μm. Hence, the electron beam heating could be more effective than laser beam heating.

Table 1. Results from experiment - heating with laser beam. The modulation frequency is 723 Hz. $\Delta R/R$ is the relative change in the reflectance.

Sample	$\Delta R/R$, μV/V	Detector gain, dB
Si	25	60
InP	10	50
InP	7	50
InP	20	60
InP	16	60

Good isolation from mechanical vibrations and better spectral isolation of the two laser beams are very important for this experiment to achieve stable results.

4. Conclusion

The possibility of combining two complementary experimental techniques – ellipsometry and scanning electron microscopy was demonstrated by theoretical calculations and simulations. The simulations show that the electron beam induces changes in the refractive index can be possibly detected experimentally by optical methods and in particular, by ellipsometry. Optimal conditions for this were estimated. Potential applications of the combined method are examining the thermal and electronic properties of micro-objects. Another potential application is using the optical channel of information for imaging.

Acknowledgments

This paper was presented at INERA WORKSHOP "Transition Metal Oxide Thin Films-functional Layers in "Smart windows" and Water Splitting devices: Technology and Optoelectronic properties", 4 – 6 September, 2014 Varna, Bulgaria. The workshop was supported by FP7 Project REGPOT 316309 INERA and organized by the Institute of Solid State Physics at Bulgarian Academy of Science.

References

[1] Fujiwara H 2007 *Spectroscopic ellipsometry, Principles and applications* John Wiley &Sons Ltd
[2] Goodhew P J, Humphreys J and Beanland R. 2001 *Electron microscopy and analysis*. third ed Taylor & Francis
[3] Song H Z, Li Y X, Zeng H R, Ma L, Wu G H and Hui S X 2008 *App. Phys.A* **92** 309
[4] Bennett B R, Soref R A, del Alamo J A. 1990 *IEEE J. Quantum Electr.* **26** 113
[5] Chan M C Y, Kwok P C K and Li E H 1998 *IEEE J. Selected Topics Quantum Electron.* **4** 685
[6] Kovalskiy A, Neilson J R, Miller A C, Miller F C, Vlcek M and Jain H. 2008 *Thin Solid Films* **516** 7511
[7] Sadagopan T, Choi S-J, Sang Jun C, Djordjev K and Dapkus P D 2005 *IEEE Photonics Techn. Lett.* **17** 414
[8] Sundari S T, Srinivasu K, Dash S and Tyagi A K 2013 *Solid State Commun.* **167** 36
[9] Lee A S and Norris P M. 1997 *Rev Sci Instrum.* **68** 1307
[10] Gillen G D, DiRocco C, Powers P and Guha S 2008 *Appl. Optics* **47** 164
[11] Russev S C, Drolet J P and Ducharme D 1998 *App. Optics* **37** 5912
[12] Farzaneh M, Maize K, Luerssen D, Summers J A, Mayer P M and Raad P E 2009 *J. Phys.D-* **42** 143001
[13] Azzam R and Bashara N 1977 *Ellipsometry and polarized light* North-Holland Publishing Company
[14] Paddock C A and Eesley G L 1986 *J. Appl. Phys.* **60** 285

AUTHOR INDEX